The Gas Mileage Bible

**How to Save Hundreds of Dollars a Year on Fuel
By Improving Gas Mileage and Performance
In your Gas Guzzling Car, Truck, Van or SUV
Without Expensive Gimmicks or Gadgets**

By
Kenny Joines
&
Ron Hollenbeck

A

Book

First Edition, Rev 2.35

Copyright © 2006 by Kenny Joines & Ron Hollenbeck

All rights reserved. No part of this book shall be reproduced or transmitted in any form or by any means, electronic, mechanical, magnetic, photographic including photocopying, recording or by any information storage and retrieval system, without prior written permission of the publisher. No patent liability is assumed with respect to the use of the information contained herein. Although every precaution has been taken in the preparation of this book, the publisher and author assume no responsibility for errors or omissions. Neither is any liability assumed for damages resulting from the use of the information contained herein.

ISBN 0-7414-3059-2

Published by:

INFI∞ITY
PUBLISHING.COM

1094 New DeHaven Street, Suite 100
West Conshohocken, PA 19428-2713
Info@buybooksontheweb.com
www.buybooksontheweb.com
Toll-free (877) BUY BOOK
Local Phone (610) 941-9999
Fax (610) 941-9959

Printed in the United States of America

Printed on Recycled Paper

Published May 2006

Contents

Part One   Overview **1**

Chapter 1:   Introduction   **1**

About the Book 1
Assumptions about You 4
About the Authors 6
Icons Used in this Book 8

**Chapter 2:   Optimizing your car for fuel
efficiency and performance**   **11**

So you drive a gas guzzler 11
Why worry about getting better gas mileage? 12
Benefits to optimizing your vehicle 13
Hybrid vehicles and alternative fuels are expensive and not readily
    available 19
What the auto dealers and the government are NOT telling you 20
Here's what you'll learn that you won't hear anywhere else 26
What is the purpose of your vehicle and how do you drive it? 27
The LED Method of Optimizing your Vehicle 29
The Three Levels of Optimization 30
The Action Time Frame 31
How do you measure and monitor your gains? 32
Before You Make any Changes to Your Vehicle 35

Part Two The LED Method for Getting Better Gas Mileage and Saving Money   37

Chapter 3: Preventing Energy Losses that Drain your Vehicle of Power and Increase Fuel Consumption   39

Types of Losses 39
Reducing Friction Losses 40
Tips and Techniques for Reducing Friction Losses 45
Reducing Mechanical Load Losses 53
Tips and Techniques for Reducing Mechanical Load Losses 64
Reducing Heat and Noise Losses 71
Tips and Techniques for Reducing Heat and Noise Losses 73
Energy Losses – The BIG Picture 73

Chapter 4: Optimizing Engine Efficiency   75

Internal Combustion Basics and Optimum Tuning 75
Optimizing Air Delivery 79
Tips and Techniques for Optimizing Air Delivery 81
Optimizing Fuel Quality and Delivery 87
Tips and Techniques for Optimizing Fuel Quality and Delivery 91
Optimizing Fuel/Air Combustion, Ignition and Timing 98
Tips and Techniques for Optimizing Combustion, Ignition and Timing 106
Optimizing Engine Exhaust 110
Tips and Techniques for Optimizing Engine Exhaust 118
Optimizing Lubrication and Cooling 123
Optimizing Mechanical Health and Tuning 124
Tips and Techniques for Optimizing Mechanical Health and Tuning 126

Chapter 5: Alternative Fuels and Our Environment   129

Beyond gasoline or diesel – alternative fuels 129

Diesel and Bio-Diesel 131
Other Alternatives 133
Our Environment 134

Chapter 6: Driving Technique and Energy Management — 135

Driving Technique Can Improve Your Gains or Take Them Away! 135
Energy Management 137
Bad Driving Behavior 141
Good Driving Habits 143

Chapter 7: Putting It All Together — 161

Recap - LED Method for Getting Better Gas 161
The Psychology of Energy Management 163

Part Three Additional Information and Resources — 167

Factors Affecting Gas Mileage (Other than the Energy Content of Fuel) 167
Learn more about saving money and getting better gas mileage 169
Other books and products available 169

Part Four Index of Tips and Techniques — 170

Tips for Reducing Losses 170
Tips for Increasing Efficiency 171
Tips for Driving Technique 172

Kenny Joines & Ron Hollenbeck

Fellow Drivers,

Here is the stuff our attorney says we need to include. Let it be known that we are giving everyone the straight scoop!

Kenny and Ron

Limits of Liability / Disclaimer of Warranty

The information contained in this material (including but not limited to any manuals, CDs, recordings, MP3s or other content in any format) is based on sources and information reasonably believed to be accurate as of the time it was recorded or created. However, this material deals with topics that are constantly changing and are subject to ongoing changes RELATED TO TECHNOLOGY AND THE MARKETPLACE AS WELL AS LEGAL AND RELATED COMPLIANCE ISSUES. Therefore, the completeness and current accuracy of the materials cannot be guaranteed. These materials do not constitute legal, compliance, financial, tax, accounting or related advice.

The end user of this information should therefore use the contents of this book and the materials as a general guideline and not as the ultimate source of current information and when appropriate the user should consult their own professional advisors.

Any case studies, examples, illustrations are not intended to guarantee, or to imply that the user will achieve similar results. In fact, your results may vary significantly and factors such as your vehicle, environment, driving style and many other circumstances may and will cause results to vary.

THE INFORMATION PROVIDED IN THIS PRODUCT IS SOLD AND PROVIDED ON AN "AS IS" BASIS WITHOUT ANY EXPRESS OR IMPLIED WARRANTIES OF ANY KIND WHETHER WARRANTIES FOR A PARTICULAR PURPOSE OR OTHER WARRANTY except as may be specifically set forth in the materials. IN PARTICULAR, THE SELLER AND AUTHORS OF THE PRODUCT AND MATERIALS DOES NOT WARRANT THAT ANY OF THE INFORMATION WILL PRODUCE A PARTICULAR ECONOMIC RESULT OR THAT IT WILL BE SUCCESSFUL IN CREATING PARTICULAR FUEL ECONOMY OR PERFORMANCE RESULTS. THOSE RESULTS ARE YOUR RESPONSIBILITY AS THE END USER OF THE PRODUCT. IN PARTICULAR, SELLER OR AUTHORS SHALL NOT BE LIABLE TO USER OR ANY OTHER PARTY FOR ANY DAMAGES, OR COSTS, OF ANY CHARACTER INCLUDING BUT NOT LIMITED TO DIRECT OR INDIRECT, CONSEQUENTIAL, SPECIAL, INCIDENTAL OR OTHER COSTS OR DAMAGES, IN EXCESS OF THE PURCHASE PRICE OF THE PRODUCT. THESE LIMITATIONS MAY BE AFFECTED BY THE LAWS OF PARTICULAR STATES AND JURISDICTIONS AND AS SUCH MAY BE APPLIED IN A DIFFERENT MANNER TO A PARTICULAR USER.

This book contains material protected under International and Federal Copyright Laws and Treaties. Any unauthorized reprint or use of this material is prohibited.

The Gas Mileage Bible

∞ ∞ ∞

For our lovely wives,
Gretchen and Kelley,
whose love and support
made this book possible.

And for all of our beautiful children,
who remind us daily of what is important:
Kim, Chris, Courtney, Jessie, Brittany,
Zack, Kolton, Tatum, and Tabitha.

We love you!

∞ ∞ ∞

Kenny Joines & Ron Hollenbeck

Part One   Overview

Chapter 1: Introduction

About the Book

The purpose of this book is to help you save money on fuel for your vehicle.

Gas prices are forecast (as of May 2006) to exceed $4.00 a gallon by end of summer. Some more extreme forecasts are predicting $6.00 a gallon within six months. There is a growing consensus we won't see any relief from these prices for at least two or three years.

What that means to us is our traditionally cheap gasoline will only become more and more expensive over the next several years. We don't honestly believe we will be going back to the good old days of cheap fuel.

We don't know how all this will play out, but one thing is for sure - we all want to save money on gas. People on a tight budget will quickly run out of options. Most people must find a way to make ends meet while paying double for gas this summer what they paid only months ago.

Our passion is to put everything we've learned throughout our lives and all our research into one place so we can share it with you. This is a work of love for us. We hope you find great value in the information you find in this book. We'd like you to start using what you learn to save money at the pump and put your savings back into your household budget where it can be used for other important things. .

While saving money, you will also be using much less gasoline or diesel fuel, which are derived from non-renewable natural resources. Globally, we are very near a tipping point where it will become increasingly difficult and expensive to extract oil from our earth. Anything we can do to conserve oil resources until alternative fuels have become common will help us all.

Using less fuel and burning that which we use more efficiently will make a big difference in the volume of pollutants and greenhouse gases being dumped into our atmosphere.

Scientists have proven that our increasingly unstable weather patterns, melting of our ice caps and other global phenomenon are directly related to global warming and the effect of greenhouse gases.

We hope to educate you and to provide you options for improving fuel economy and efficiency in your vehicle. We teach you three factors that affect your gas mileage. We tell you how they relate to specific components of your vehicle and to your driving habits. And, finally, we give you specific tips and techniques you can implement yourself or entrust to a handy friend or neighborhood mechanic.

This book is about energy. We are rapidly consuming one form of energy we have come to depend on as reliable, bountiful and cheap – and that, of course, is oil. If we can help even a few of the millions of drivers who waste hundreds of millions of gallons of fuel every year because of inefficient vehicles and bad driving habits, then we have gained a little time to learn how to harness other forms of energy. It is just a matter of time (some experts predict within a few years) that oil will become so hard to get and so expensive, that we will be forced to make drastic changes, not just in our cars and our driving habits, but our entire way of life.

Think about it. In one way or another, most things in our lives depend on oil or gas – electricity, vehicles, farming, manufacturing, buildings. It will take a major shift in how we harness energy for us to maintain our forward progress as a civilization.

So, this book is focused on conserving energy. It is about doing what we can do with what we have to make a difference. We believe that while you are making a difference in your pocketbook, you are also doing your share to conserve our limited oil resources, and you are putting fewer pollutants and harmful gases into our air.

This book is not intended as a maintenance manual. Nor is it intended to provide step by step instructions for implementing solutions we recommend. Our goal is to present you with information and a range of possibilities for improving your fuel economy and letting you take it from there.

There are other side benefits to improving your fuel mileage. Your car runs more efficiently so you gain additional power and performance. Most drivers who follow advice in this book feel noticeably improved acceleration and drivability. Those who haul or tow will find additional power they didn't know they had. Another benefit is that your vehicle will last longer and run with fewer maintenance problems if you follow our advice and keep your vehicle tuned and operating efficiently.

We hope you will learn a little bit about cars, and improving your gas mileage, vehicle performance and environmental friendliness by reading this book. We hope you like our ideas regarding energy management, think about them often and incorporate those ideas into your daily drive. Let your new energy management concept provide context around which to view all your driving habits, as they play such a vital role in the whole fuel economy equation.

Our greatest hope is that you will have a fresh, new perspective on driving that will carry over into your entire life; that you feel a new pride of ownership, you know how to achieve maximum efficiency and you become the safest driver possible, truly enjoying your driving experience.

And, of course, we'd like you to save lots of money at the pump!

Assumptions about You

We've made some assumptions about you, even though we haven't met yet. (Isn't it true that a stranger is just a friend we haven't met yet?) We assume you own a vehicle, probably a big one, that doesn't get great gas mileage. You are tired of paying so much for gas; spending much more than you care to each week filling up your tank - as much as $100 or more for some of you with trucks or SUV's. You are looking for a better alternative.

Knowing this, we expect you are just as afraid as we are about near term driving costs.

We are also assuming you don't have plans to trade your vehicle in on a different one – that for whatever reason, you are going to keep your gas guzzler for at least a couple more years and make the best of high fuel prices.

You may have a specific need for your larger vehicle. Maybe you need it for work or for hauling your soccer team around. Many, many people have become dependent on their full size vehicles.

You may have tried to sell or trade-in your truck or SUV, only to find out you owe more than your dealership will pay you. This is called being "upside down" on your financing. It's likely that hundreds of thousands, if not millions, of people will be stuck with their larger vehicles because of gas price increases.

So whatever your reasons, you've decided to stick it out, keep your truck or SUV and figure out how to get the best possible mileage.

We don't know if you are handy with cars, or tools, or even if you can walk and chew bubble gum at the same time. But that's ok. In this book, we are not going to get into precise details of how to perform the recommended tips and techniques. We assume you know enough to perform basic "no-tech" and even many "low-tech" solutions presented here. We also assume you have a relationship with a mechanic or a knowledgeable friend you trust who can help you with things you don't know how to do.

About the Authors

Each of us (Kenny and Ron) grew up around cars. We both built our first hot rods in high school and spent our entire lives being car (and truck) junkies.

Kenny's Dad is an aerospace engineer whose life hobby is building, driving and showing cars. He routinely wins shows around the Dallas - Fort Worth area with his vehicles. Ron's Grandfather was a mechanic who taught him and his brother, Gary, how to build their first hot rod – a shiny blue metallic '57 Chevy.

Kenny and Ron's love for automobiles has extended into their adult lives, where they constantly tweak, test and fiddle with their current vehicles.

A key concept in this book is that of energy management. Much of what we discuss centers around managing your vehicle's energy, whether it be energy produced by your engine or momentum you have driving down the road. Each of the authors received "special training" in energy management early in our careers that carried over and became the substance of what is now this book.

Kenny's view of energy management was through the eyes of a hard-hitting and fast-running football player. He achieved great success on the field even without the dominating size or speed of some of his teammates. He learned that success on the field is dictated by your ability to first understand the energy you have, the energy your opponent has, and then know how to focus and apply your energy at any given moment.

Manage your energy and the energy of your face-to-face opponent. If you use your energy effectively and leverage that energy to get past, block or tackle your opponent, while denying him the ability to use his energy to its fullest, you are successful on that play. To be successful on the field, you must be fully aware of the nuances and details around you, anticipate what's going to happen next before your opponent does, and apply vision, focus and technique. If you don't do this well, you go down, and probably pretty hard. Kenny found that the more competitive your environment and the tougher your objective, the higher the degree of preparedness, focus and situational awareness you need to survive, let alone thrive.

Ron's view of energy management was through the eyes of a Naval Flight Officer, flying Navy jets and landing on aircraft carriers. He learned early in flight school the importance of managing the airplane's energy. He went through air combat maneuvering school (dog-fighting) and learned an entire art and science around managing energy. That's the main thing dog-fighting aviators really worry about – how to sustain their energy for the longest amount of time. There is an incredibly detailed science around managing momentum, converting kinetic energy (speed or momentum) into potential energy (altitude) and back again. Every aircraft has a fixed amount of energy available to it. By maximizing use of energy and forcing your opponent to make a mistake and lose some of his, aviators learn how to overcome their opponent, get "on their six" *("on their six" is an aviator term meaning to get directly behind the other aircraft. It correlates to the six o'clock position on the face of the clock where the opponent's aircraft is in the center)* and shoot him down. He with the most energy at the end wins.

Driving your vehicle efficiently is really about maximizing energy produced by your engine through the mechanical operation and maintenance of your vehicle and through wise and conservative driving habits. We hope you enjoy the book and learn how to manage energy while saving money along the way.

Icons Used in this Book

General Information

 Warning. This note is related to safety or something that will keep you out of trouble. If you don't heed the warning, it could lead to injury, lost time or money.

 Anecdote. Just an interesting factoid based on real life experience. It might make you laugh, it might make you cry, but it will always illustrate a point in the book.

 Tip. Some advice from us to you to keep you focused.

Solution Type

 Solution for reducing **Losses**, like friction or air resistance, one of three factors that influence your gas mileage.

 Solution for increasing engine **Efficiency,** one of three factors that influence your gas mileage.

 Solution for improved **Driving** technique, one of three factors that influence your gas mileage.

Solution Complexity

  No-Tech   No tools or technology involved in this solution

  Low-Tech   Simple tools or low sophistication involved in this solution

  High-Tech   Advanced tools or technology; more sophisticated solution

Solution Time Frame

  Immediate   Things you can do today

  Short-term   Things you can do within thirty days

  Long-term   Things you can do as you have time or budget

Kenny Joines & Ron Hollenbeck

Chapter 2: Optimizing your car for fuel efficiency and performance

So you drive a gas guzzler...

Necessity is the mother of all inventions. So you have a gas guzzler and you need it! Maybe you work in construction and need that pickup truck, or maybe you are a soccer Mom who needs the big Suburban or Expedition to get all the kids, friends, pets and gear to and from wherever it is they need to go.

Or maybe you just like the size, expanse and road-presence a big SUV provides. Some of us get to choose what we drive, and we drive our gas guzzler by choice.

In Texas, many of us drive trucks because, well, that's what we drive in Texas! The bigger and taller, the better. If only we could get a little better gas mileage and spend a little less time and money at the gas pump....

There are many types of gas guzzlers – large cars, sports-cars, trucks, vans and SUV's. Many Americans drive these vehicles for their own reasons and by their own choice.

But all of us who own gas guzzlers have this in common – we want to pay less at the gas pump and spend less time maintaining and keeping our vehicles for the privilege of driving these beautiful and functional machines.

Why worry about getting better gas mileage?

Oil prices are on their way up again. We thought we had seen the worst of it during late summer and fall of 2005 after Hurricane Katrina did so much damage to the Gulf Coast and oil production there. Then, we were happy to see gas prices drop once again to $2 a gallon. And we were ecstatic! Who would have thought we'd be so happy to see $2 a gallon gas?

And here it is, going up again in the spring and summer of 2006.

Well, we have news for you. *Get used to it*! Oil prices are going back up and so will your gas bill. Expectations are already being set for as much as $4.50 a gallon prices this year. The rash of hurricane weather events of 2005, political unrest, fear of terrorism, the Iranian nuclear threat, increased regulatory pressure to provide cleaner fuel, tight supplies, and summer travel season are just a few reasons and catalysts for spiraling gas prices – and for us to publish this information.

So with oil and gas prices skyrocketing, what other recourse do we have? For some people, buying smaller or hybrid vehicles may be the answer. If this is an option for you, and you own a larger SUV or truck, we recommend you pursue it quickly before resale values for these vehicles drop even more than they are now. Be aware, as fuel prices go up and more people decide to sell their SUV's, the possibility of getting out from under your gas guzzler and not losing your shirt goes markedly down. Just look at your local Saturday newspaper ads and you'll see the signs - discounts galore for the gas guzzler of your choice in full page glory. As gas prices increase and the market gets flooded with big vehicles, your options will become increasingly more limited.

A few people may be able to share rides, use a bike or even walk to work. A friend of ours has given up car ownership completely. Whenever he needs to drive somewhere, he rents a car. In between, he walks, catches a ride with someone or rides his bike. He finds he saves money and headaches by not paying for vehicle financing, insurance, maintenance, gas and all the other expense that come with car ownership.

But this may not work for very many of us. For the rest of us, we need to learn how to optimize our vehicles for better fuel economy and savings at the pump.

We believe most vehicles can easily get from *10 to 30% increase* in fuel economy using the tips and techniques in this book. A serious gas guzzler could possibly achieve as much as *40% improvement or more*!

The purpose of this book is to teach you how to optimize fuel efficiency of your vehicle and save money. That way, you won't be forced to buy a smaller vehicle that may not serve your needs, or be forced into some other drastic measure, like abandoning your vehicle one day when you can no longer afford to fill it up.

Benefits to optimizing your vehicle

Saves you Money

Efficiency can be a beautiful thing. What if you could get 10 to 30% improvement in your fuel mileage? How much could that save you each year?

For example, if you typically get 12 miles to the gallon (MPG) and drive 15,000 miles per year, you could **save over**

$900 per year if you could improve by 5 miles per gallon, up to 17 MPG, an improvement of 40%.

Following are some examples of real world savings that are possible. Just look on each table for your current average gas mileage and then read straight across to see what your savings would be at 10%, 20% and 30% improvement!

Savings for someone who drives 15,000 miles per year, spends $4.00 per gallon for gas, and gets **10% improvement**:

Current Average MPG	Current Gallons/Year	Current $/Year	Improved Average MPG	Improved Gallons/Year	Improved $/Year	Annual Gas Savings (Gal)	Annual $ Savings	Monthly Savings
10	1,500	$6,000	11	1,364	$5,455	136	$545	$45
14	1,071	$4,286	15.4	974	$3,896	97	$390	$32
18	833	$3,333	19.8	758	$3,030	76	$303	$25
22	682	$2,727	24.2	620	$2,479	62	$248	$21

Savings for someone who drives 15,000 miles per year, spends $4.00 per gallon for gas, and gets **20% improvement**:

Current Average MPG	Current Gallons/Year	Current $/Year	Improved Average MPG	Improved Gallons/Year	Improved $/Year	Annual Gas Savings (Gal)	Annual $ Savings	Monthly Savings
10	1,500	$6,000	12	1,250	$5,000	250	$1,000	$83
14	1,071	$4,286	16.8	893	$3,571	179	$714	$60
18	833	$3,333	21.6	694	$2,778	139	$556	$46
22	682	$2,727	26.4	568	$2,273	114	$455	$38

Savings for someone who drives 15,000 miles per year, spends $4.00 per gallon for gas, and gets **30% improvement**:

Current Average MPG	Current Gallons/Year	Current $/Year	Improved Average MPG	Improved Gallons/Year	Improved $/Year	Annual Gas Savings (Gal)	Annual $ Savings	Monthly Savings
10	1,500	$6,000	13	1,154	$4,615	346	$1,385	$115
14	1,071	$4,286	18.2	824	$3,297	247	$989	$82
18	833	$3,333	23.4	641	$2,564	192	$769	$64
22	682	$2,727	28.6	524	$2,098	157	$629	$52

We have a nifty little calculator that tells you how much you can save by improving your fuel efficiency. It helps you calculate how much you spend now and how much you will spend with improvements you make. You can find the calculator on our web site at:

www.vitalshift.com

These savings are entirely possible. The actual amount you will save depends on your vehicle, your environment, your driving habits and your willingness to implement the tips and techniques in this book.

Ask yourself this question: "What can I do with several hundred more dollars this year?"

Saving money over the entire lifespan of your vehicle will be longer if you apply some or all of the tips and techniques provided you in this book. In the first example above you would save more than $4,500 over five years!

Since your engine is working more efficiently, it needn't work as hard to achieve the same results. Because of that, you will likely have a noticeable improvement in performance.

Another result of using our tips and techniques is that the moving parts have less friction. Friction is an enemy and barrier to getting better gas mileage. Less friction between moving parts means greater efficiency. Moving parts don't wear out as fast, meaning your vehicle requires less maintenance and it lasts longer!

Hmmm. Lower cost, greater performance, lasts longer. Sounds like a good deal to us!

Protects the Environment

Depletion of our nation's and our world's oil reserves is a topic we've all heard about. The big debate is how long our oil reserves will last us. There have been predictions over the years saying our oil reserves may be running out and environmentalists have raised the flag concerning this urgent situation.

In reality, the public does not really know how long our oil will last. It makes sense to do something about it though.

We graduate hundreds of thousands of new drivers each year and the life expectancy of adults in the U.S. continues to climb. What that means is more and more cars are taking to the highways each year.

Most of the durable goods and GNP (Gross National Product) of the US are tethered directly to automobile, light truck, heavy truck, new home, and other durable goods manufacture, distribution, and consumption.

Now, throw in the massive consumption of durable goods and resources by newly developing countries in the Far-East, Middle-East and Southern Hemisphere. These new energy consumers, numbering in the billions, will cause a never-before-seen increase in demand for energy of all flavors. They will require more electricity, gasoline and diesel fuel, heating and cooling, factories, and farming equipment – all of which is dependent on oil. When added into the global equation for supply and demand of energy, the demand in the near future will continue to outpace the supply, leading toward a natural trend for prices to increase, not decrease.

Following is a diagram from the Department of Energy. It depicts the sources and uses of energy consumed in the United States in quadrillions of BTU's:

The Gas Mileage Bible

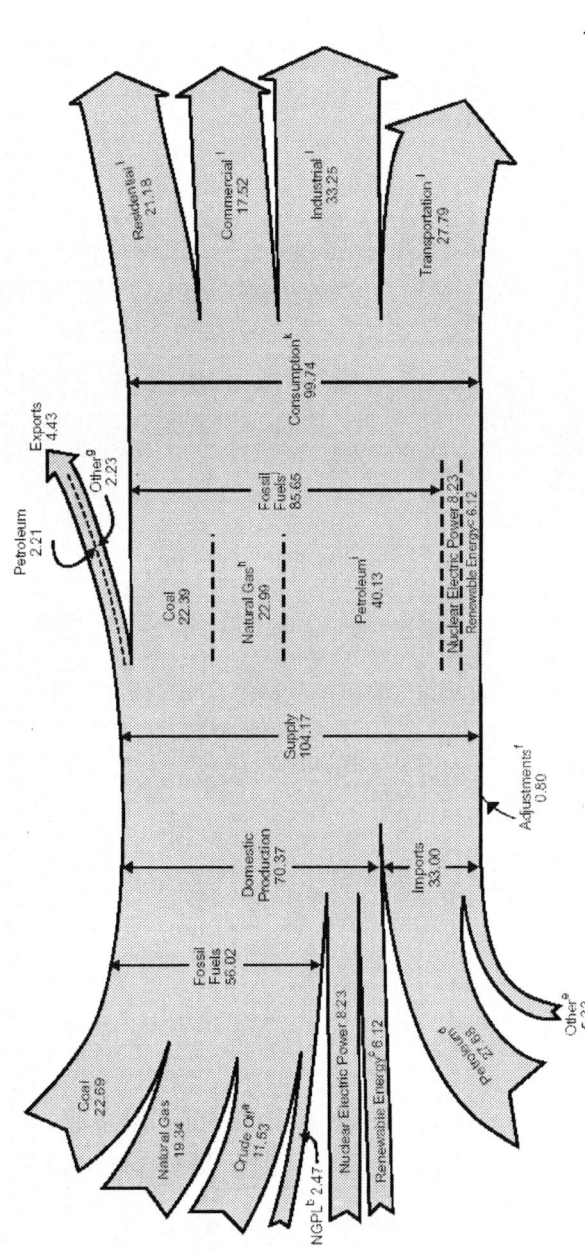

As you can see, our usage is not going to go down; it is destined to go up year after year!

It is the simple law of supply and demand. Even the possibility for decreased energy supply, as was seen in July through October of 2005, can drive an overnight increase in fuel, gasoline and diesel prices from 20% to 40% at the gas pump!

Greenhouse gases are another high publicity topic in the news. This is the effect of increasing global temperatures due to carbon dioxide emissions, largely due to automobile exhaust.

Reducing the amount of gas you use would help both of these situations. By improving your fuel economy and implementing our tried and true tips and techniques, you will directly reduce the amount of gas you use, thereby slowing the depletion of our natural resources. Your vehicle will also be more efficient, and will emit fewer greenhouse gases and other pollutants into the environment.

Helps our Country

Dependence on foreign oil has led to many political, economic and security issues for our great country. By reducing the amount of gas you use, you will be contributing to an improved and more secure economic position of our country.

We'll be less dependent on the oil from other countries (refer back to the diagram on page 15) where we may be forced to pay exorbitant prices to get the fuel we are addicted to. More than half of the oil we need for gasoline is imported from other countries. The United States spends over $3 Billion dollars a week for all the oil we import. Do you suppose we would see fewer political and security threats if we didn't depend on others for so much for our fuel? And, what if all that money were spent on our own economy, and not on making oil producing countries even wealthier?

There have been many skirmishes and wars fought over oil. While there are other, longer term ways of reducing our dependence on oil, such as increased use of alternative fuels and hybrid vehicles, the immediate solution *each* of us can help with is to reduce our own fuel consumption.

Hybrid vehicles and alternative fuels are expensive and not readily available

Availability of new hybrid cars, alternative fuels and vehicles that use those fuels is still low. Many hybrid dealerships have waiting lists of several weeks, or even months. Even though hybrid vehicles are wonderful and are probably the wave of the future, they are not as efficient and inexpensive as many would like.

In many parts of the country, alternative fuels are not available even when drivers have vehicles that accept them. By working on improving your gas guzzler fuel economy now, even if you intend to move to hybrid or alternative technologies in the future, you are saving money and doing yourself and your fellow citizens a big favor.

What the auto dealers and the government are NOT telling you

What You Get in the Public Domain

Common-sense advice regarding how to improve fuel economy abounds in the press, in magazine articles, on government web sites – in many places you care to look. And, most of that advice is sound if not a little obvious. Look how most sources will tell you to save gas:

- Drive less by walking, biking, taking the bus or sharing rides (duhh!).
- Keep your car tuned up.
- Keep your tires inflated.
- Change your oil frequently and use the right kind of oil.
- Drive slower.
- Don't accelerate as hard.
- Anticipate stops and brake more slowly.
- Unload excess weight from your vehicle.
- Don't use your air conditioning as much and roll your windows down.
- Don't leave your car running for long periods when stopped.

This is good advice – most of it. But you get only part of the story. You can do many other things. In this book, we have collected the greatest number of tips and techniques for improving your gas guzzler's fuel economy you will find anywhere. And, we've wrapped them up in a thorough discussion of how your vehicle works and why these tips and techniques will deliver for you.

We also tell you what NOT to do. Driving the wrong way can totally override all the mechanical gains you made by optimizing your vehicle – did you know that? Certain poorly operating components in your vehicle (including your heavy right foot on the accelerator!) can cost you as much as 40% in the gas mileage you are getting now!

Most of you MUST drive. You don't have an option to walk or ride the bus. What if you live outside of town and it's too far to walk? There are no buses and no one with whom to ride. You need to travel a busy highway and the old bicycle is just not going to cut it.

Unfortunately, because of the way towns and cities are designed in the U.S., and because our public transportation systems are completely inadequate (compared to other countries), most Americans are compelled to own their own cars and drive every day to work.

Some of you could probably follow all of this advice. That is if you live in the city within walking or biking distance of work or school. Or if you live near a bus route or have coworkers with whom to carpool.

Some of the public domain gas saving advice just plain doesn't apply to many of you, due to the nature of your driving situation, where you live and your occupation.

So, there are obvious, but maybe not so applicable methods for improving our gas mileage. That's what we get from most public sources.

You also see all the over-hyped gadgets and additives that are supposed to miraculously make your car go 100 miles on a gallon of gas; or run off of water (maybe not too far in the future – but not today!); or fly like Chitty Chitty Bang Bang.

Who knows if all that stuff really works? Beyond a few sound scientific principles that have been proven over time to work, and have been tested by thousands or even millions of everyday drivers like you and me, there are not very many miracle products out there.

In this book, we cut through all the noise and present you with what we know are the tried and true methods of saving gas. Our approach to dealing with all the gadgets and additives out there is to study real testing results, not some grandmother's testimonial. Our point of view is that an additive or gadget can be called effective if it has real testing results by a certified, nationally recognized testing authority, and it can show years of positive results with satisfied customers. Too many of these products have been scams that have been shut down by the Federal Trade Commission, or they are found later to have caused internal damage to the engines they were supposed to be helping.

We have personally tested many of the gadgets and additives being marketed today, and found that many of them don't work.

By following the method we point out to you, we think you will be surprised at how simple it is to really save gas. We know you are going to save some money if you strictly follow even a few of the recommended tips and techniques we offer to you.

What You Don't Hear About

What the automakers don't tell and you and they don't want you to know is that most of the vehicles rolling off the production line are *under-engineered*.

The automakers know a wide variety of drivers will purchase any given make of vehicle. These drivers will have many different uses for the vehicle, different driving habits and different levels of discipline in maintaining it. You may have an electrician, a motocross enthusiast and a caterer all decide to buy a certain pickup truck.

Do the automakers tune the vehicle to perform best in the work truck role, weekend hauling role, or the delivery role? They just don't know specifically how the vehicle may be used, and we're sure there are some weird ways people use their vehicles out there. So, to please the general buying public, the automakers must engineer the vehicles to do all things generally well. Have you ever heard of the phrase "jack of all trades and master of none"? To a large degree your vehicle has to be able to do generally well at many, many, things. While SUV's and trucks do their general mission well, they do not perform well in the area of gas mileage.

Have you seen the commercial on TV by GMC? It shows a number of their vehicles going through an assembly line. Each vehicle emerges as something different – a work truck, an ambulance, and even a CSI (the TV show) GMC Denali. The voice-over says something to the effect that GMC doesn't know how we'll use their vehicles, so they make them to suit a variety of purposes.

We couldn't have made that point better ourselves!!

The automakers *don't* know how we will use the vehicles we buy. A GMC truck used for hauling a trailer cross country should be engineered and tuned completely differently from a work truck or an ambulance.

So, they engineer the vehicle for the lowest common denominator, knowing the vehicle could be engineered to achieve better gas mileage, better performance and cleaner emissions given a particular driving environment and vehicle usage.

They don't know the specific usage patterns, or environment, or type of driver, so they calculate an "average" driver profile and tune to that.

Another thing the automakers and the government do not want you to know is the EPA mileage rating of vehicles is a laboratory rating and estimate at best, not a real world, driving in traffic with three screaming kids and a ringing cell phone rating.

The government has set up a very structured, strict and standardized methodology for testing fuel economy of vehicles using dynamometers and test equipment in a controlled laboratory. The vehicles never actually drive on the road when being tested; they are subjected to simulated highway and city driving conditions. To make matters worse they even control the temperature, humidity, and even the amount of air-flow the vehicle can consume and have access to during these tests.

The vehicle manufacturer's goal is to pass the government's EPA tests and to get the highest possible EPA city and highway numbers. They want the highest possible EPA mileage number approved to be placed on the new window sticker you looked at when you purchased your new vehicle.

Why, you ask? Because it helps them to sell more vehicles.

It is safe to say most vehicles' real world fuel economy is rarely as good as the EPA rating on the window sticker of your new vehicle.

Another regulatory requirement is that automakers pass an N.V.H. (Noise, Vibration and Harshness) test proving noise, vibration and harshness levels are within federally specified limits for all new automobiles. The two tests are often counterproductive to one another in terms of achieving pure efficiency in any given area. For example, an automaker might need to make a vehicle heavier with sound deadening material to pass the N.V.H test. A heavier vehicle is then less fuel efficient.

As you can see, automakers need to strike a careful balance in the design and engineering of your vehicle. In effect, this trade-off means vehicles are under-engineered for any specific performance attribute. They are not as fuel efficient because of N.V.H and emissions requirements. They are not as quiet or vibration-free because of the desire to keep the vehicles light for EPA fuel mileage testing.

Note

Some of the newest, high-end automobiles, particularly performance cars, are being engineered to a much higher standard of efficiency and performance and are not necessarily subject to some of the mechanical tips and techniques provided in this book.

In all cases, however, the principles regarding tune-ups, maintenance, minimizing losses, and driving in a disciplined manner will help you get better mileage in your newer, performance or precision engineered vehicle.

Here's what you'll learn that you won't hear anywhere else

Because automakers don't want to advertise that their vehicles could be tuned much more efficiently, they are not in the business of telling you HOW to do it! But that's ok, because we are here to tell you all you need to know!

We will tell you specific situations when the common wisdom is just plain wrong! We will provide additional detail and specific driving techniques for improving mileage and operational efficiency.

We will provide you specific tips and techniques to save you money, improve your fuel economy and make your car last longer. We will teach you all the areas of your vehicle that may contribute to or take away from its efficiency and performance.

We'll also show you how to avoid the numerous gimmicks and gadgets out there. We'll tell you which types of devices and chemicals work, based on sound scientific and engineering principles, and we'll tell you what to avoid.

Look at a few of the techniques you will learn:

- A system for eliminating problems that hurt your gas mileage, while improving areas that create efficiencies.
- A method for getting greater enjoyment out of your driving time, while reducing stress, and saving money.
- The most thorough collection of tips and techniques that actually have shown results for improving fuel economy available anywhere.

- How your vehicle generates energy by converting fuel to mechanical motion, and how your energy gets robbed, thereby losing efficiency, power, and fuel economy.

- Simple, yet effective driving techniques derived from the energy management techniques of fighter pilots that maximize the efficiency and gas mileage of your vehicle.

- How to maximize the efficiency of your engine so it taps the maximum amount of energy from a gallon of gas.

- How to use special types of oils and lubricants in your engine, transmission, and gear boxes that will increase your gas mileage by up to 15%. Not only that, under some circumstances, you will only need to change your oil every 10,000 miles or more. Your engine, transmission and gear boxes will last longer, and you will be using less of our natural petroleum reserves!

What is the purpose of your vehicle and how do you drive it?

Need drives many people's choice of a vehicle. Pure desire drives other's choices. Some people drive a truck because they haul or tow things and need a rugged heavy duty vehicle. Others drive trucks to be cool and have never actually hauled anything in the back of it. (Just look at the beds of certain trucks – no scratches, no dirt, no wear and tear!)

Some people drive a small economy car because they commute long distances. Others have sports cars because they want performance and the cool factor.

The underlying function of your vehicle and your own driving habits are something to keep in mind as we begin to learn ways to optimize it.

Drivers of cars come in all shapes and sizes, with all different motivations for being behind the wheel. If you are a delivery driver, you're in a hurry to make all the stops of the day, stay on schedule, and get back home at a decent hour.

As a soccer mom, you are probably interested in getting the kids to and from school, practices, and appointments in time get to the grocery store, the post office, the dry cleaners and then home to dinner and homework.

If you are a business or sales person, you may be most interested in going straight from point A to point B, allowing time to think and conduct business along the way. You may also be a rural long-haul driver or sales executive trying to make the best time, while driving hundreds of miles each day, to get from one region in the country to the next.

Along with that time, you are paid a mileage expense for each mile driven so saving money per mile is like putting money in your pocket!

Everyone operates their lives with different motivations, in different environments and at a different pace. There are an unlimited number of ways vehicles get driven to suit people's lifestyles – short or long distance, city or highway, light or heavy payload, light foot or lead foot, or rough, hilly terrain, or on smooth well kept roads.

All of these factors play a role in the quality of results you will have in optimizing your gas guzzler.

The LED Method of Optimizing your Vehicle

Systematically assessing all the areas where fuel economy can be improved is part of the *LED Method for Optimizing Your Vehicle*.

We look at *three key factors* affecting gas mileage – **LOSSES** of energy due to friction and other forces, **EFFICIENCY** of the engine itself, and **DRIVER** technique.

For each of these three areas we methodically address the parts of your vehicle and functions impacted, and provide recommendations for making that area more efficient.

 Losses — Forces such as friction and wind resistance that slow down movement and reduce the efficiency of our vehicle

 **E**fficiency — The ratio of the energy delivered by a machine (the engine) to the energy transferred, supplied, and used wisely for its operation. Another way of looking at it is how well the energy in a gallon of fuel is converted to mechanical energy in the engine.

 Driver — The person who turns, accelerates, and brakes the vehicle and the collection of driving habits used by that person to maneuver the vehicle in either an efficient or an inefficient manner

All vehicles operate under the same laws of physics. A force is applied to make a vehicle move forward (the engine) and at the same time, other forces try to hold it back, such as air resistance, friction, mechanical load and the force of gravity and other physical forces.

The efficiency of the engine in pushing the vehicle forward determines how hard it has to work, given the same level of force (losses) it has to counteract. On top of all this, the driver plays an amazingly important role in managing the potential and kinetic energy (remember high school physics?) configuration of the vehicle and the efficiency at which the vehicle operates.

We'll be addressing each of these factors in detail in Part 2 of this book.

The Three Levels of Optimization

No-tech (no technology) and **Low-tech** (low technology) solutions are our favorite methods for optimizing your Guzzler. These are methods requiring little or no technical knowledge and are relatively inexpensive to implement.

We will focus on the No-Tech and Low-Tech aspects of optimizing in this book. Some people may pursue high-tech solutions, but these either require a higher level of technical sophistication, a big fat pocketbook or both.

We will mention one or two of these high end solutions, but will leave them for the topic of another book. We will identify the level of complexity of our solutions with the following icons throughout the material in this book:

No-Tech   No tools or technology involved in this solution

 Low-Tech   Simple tools or low sophistication involved in this solution

 High-Tech Advanced tools or technology; more sophisticated solution

The Action Time Frame

Similarly, immediate, short-term and long term actions will be identified in the following manner:

 Immediate   Things you can do today

 Short-term Things you can do within thirty days

 Long-term   Things you can do as you have time or budget

How do you measure and monitor your gains?

Recordkeeping is a big part of optimizing your vehicle and maintaining high fuel efficiency. *If you don't measure it, you can't manage it.*

Awareness of how your vehicle is performing is the first step in operational efficiency.

You need to track certain things and measure the results of your efforts. Before you begin implementing any of the tips and techniques in Part two of this book, we strongly encourage you to have a good idea of what your mileage is now.

This is called "baselining" your vehicle. All that means is for three to four weeks you track the amount of miles you drive and the amount of fuel you put into your vehicle. You should be aware of the city and highway MPG you are getting and under any other special circumstances such as towing, hauling or driving in difficult environments.

If you do not know what your current mileage is, put this book down now, track your mileage for two or three full tanks of gas, and then pick up where you left off.

If you do not know how to calculate your gas mileage, here is the quick and easy way to do it:

- Fill up with gas. At that time (not before, not after) jot down the mileage on your odometer in a little notebook or in your electronic organizer.
- Drive your car until its time to fill up again

- Fill up your car and jot down the mileage on your odometer again as well as the number of gallons of gas you filled up with at that time.

- Subtract the odometer reading you had last time you filled up with fuel from this reading. That is the number of miles you have driven.

- Divide the number of miles you have driven by the number of gallons you added to your tank. This is your gas mileage or miles per gallon (M.P.G.)

- Write down your gas mileage and other notes about the driving conditions you experienced while using that last tank of fuel. You may want to write down who drove the vehicle, was it highway or city driving and so on.

We encourage you to keep an ongoing log of your fuel consumption and mileage. It's easiest to keep it in a small notebook you keep in your glove box or console. Make it a habit to write this information down each and every time you stop for gas and fill up.

Fill 'Er Up – Every Time!
This method does not work if you do not fill your vehicle up to the same level every time. You need to FILL UP THE TANK every single time. If you don't fill the tank up one time, you need to skip the monitoring until you fill up again, at which point you can restart the cycle as described above.

We (the authors) have electronic organizers (Palm Pilots) where we keep this information. You can download lots of free or inexpensive programs off the web that include mileage tracking databases.

A couple of them are listed here:

- "SmartList" for Palm – this is a database application with many pre-built databases you can download, including "Fuel Record" which is a very nice and simple fuel tracking program. You just put your odometer reading, gallons and cost in and it calculates miles per gallon and cost per mile. It also tracks where you filled up and allows you to track multiple vehicles if you choose.

- Other Palm-based fuel, travel and productivity tools can be found at:

http://www.tinystocks.com/highway.html

http://www.palmone.com

You can also use Microsoft Excel on your PC to track your mileage and do all the calculations for you. A nice thing about using Excel is that you can chart and graph your data and start to visually see trends that you may not notice when you are looking at a list of numbers.

We recommend that all of you looking to reduce your fuel consumption keep this log and analyze it from time to time to track your vehicle's efficiency.

If you see a sudden dip in mileage and nothing else has changed, your early warning system may have just announced the early stages of a mechanical or tune-up problem with your vehicle.

Keep a Log

Keep a maintenance log of your vehicle that includes all oil changes, tune-ups, tire maintenance, brake jobs, shop visits, and so on. The idea is to keep a record of anything that you do to the vehicle. You may not realize it, but having that information available is likely to increase the resale value of your vehicle later on. Prospective buyers are willing to pay a premium for a vehicle that has records to show that it has been well maintained and is in top notch condition.

Before You Make any Changes to Your Vehicle

Before you get started implementing some of the tips or techniques in this book requiring physical modifications to your vehicle, check with your owner's manual and your dealer for the following:

- Make sure planned modifications will not affect your warranty – we will try to identify which modifications may impact your warranty, but there is a lot of variance between dealers and we won't always be able to say for sure.

- Make sure your vehicle has not already been engineered with a capability we are suggesting you implement. For example, some new high end cars come standard with higher flow multi-staged air intake systems. You wouldn't want to rip one of those out to replace it with an aftermarket system!

- Check out the web site and other marketing materials for any product you choose to use on your vehicle. You want to look for certification that their product has been approved for use on cars in all 50 states. Also, look for endorsements by reputable organizations. Please be careful. There are lots of shady products out there.

Do Your Homework – *Before* You Modify

Do your homework before making any physical changes to your vehicle. See the points above to learn what is feasible and what is not feasible for your vehicle.

Part Two The LED Method for Getting Better Gas Mileage and Saving Money

Kenny Joines & Ron Hollenbeck

Chapter 3: Preventing Energy Losses that Drain your Vehicle of Power and Increase Fuel Consumption

Types of Losses

Energy management is the key to fuel efficiency. Prevent your vehicle from losing too much of the energy produced by your engine to friction and other forces, and you have made a big step in creating a more fuel efficient vehicle.

There are inherent and ever-present forces always draining energy from your vehicle. The three basic types of losses are as follows:

Friction — wind resistance, rolling resistance, mechanical friction (two parts moving against each other), and hydraulic friction (force required to push or pump a liquid)

Mechanical Load — the combination of losses incurred by the engine due to powering accessories such as air conditioning, electrical, and power steering as well as vehicle weight and payload

Heat & Noise — loss of energy in the form of heat loss or vibrational noise

Always being aware of and reducing any of these energy draining forces will allow more of the energy produced by the engine to be transferred directly to turn the wheels and propel your vehicle more efficiently.

An increased level of any of these draining forces adds to the burden your engine must carry as it attempts to power your vehicle. For example, a heavier payload causes the engine to work harder under load, just as tires with low pressure create extra resistance and make the engine work harder.

This section will address each of these types of losses and offer recommendations for reducing or even eliminating some sources of the ever-present losses that drain energy from your vehicle.

Reducing Friction Losses

Wind and Rolling Resistance

Aerodynamics plays a very important part in the efficiency and fuel economy of your vehicle. You have noticed that sports cars are sleek and streamlined, and for a reason. The car's shape is more finely tuned to allow air to pass more cleanly over these vehicles with a minimal amount of turbulence.

The smoother and better managed the air flow over a surface, the less turbulence is formed, and the less air resistance or friction that needs to be constantly overcome. Of course, since you probably have a gas guzzler, it is probably not as streamlined or aerodynamic as a sports car. In fact, most SUV's and trucks can be more closely called "rolling-bricks!"

Larger vehicles are made with another purpose in mind other than for low wind resistance. Many trucks and vans are box-like in shape to accommodate cargo and big engines. The same goes for SUV's which are generally designed to carry many passengers and large-volumes of cargo as well.

These vehicles are not very aerodynamic and as a result tend to have bigger engines and more horsepower to overcome the increased air resistance. In fact, engineers will tell you the drag or wind resistance of a vehicle is *proportional to the square of the speed.* In other words, if you double your speed, you quadruple the aerodynamic drag you must overcome – *that's four times more resistance*!!

And here's the kicker – the power needed to overcome that drag increases proportionally to the cube of the speed. So, if you double your vehicle speed, *you need eight times the horsepower to overcome four times the drag.*

Here's another example. If you increase your speed from 40 to 60 miles per hour, that's 1.5 times faster. You get 2.25 times more resistance (1.5 x 1.5 = 2.25). Your engine needs to come up with 3.4 times the horsepower (1.5 x 1.5 x 1.5) to keep the vehicle moving along at 60 than it did at 40 miles per hour.

You now start to get a feeling for why trucks and SUV's require such big engines, and why mileage experts tell you to drive slower to conserve fuel.

Rolling resistance is exactly what it sounds like. It is the resistance of the rubber tires rolling over the road surface, combined with the friction of the brakes and wheel bearings.

Amazingly, *one third of the total energy output* of the engine is consumed by rolling resistance. Other factors playing into rolling resistance are the pressure, quality and type of tires, your alignment and the weight of your vehicle.

New radial tires have the best of both worlds with a comfortable ride as well as much lower rolling resistance than the older bias ply tires - *almost double the efficiency*!

Taller and fatter tires obviously offer higher wind resistance than smaller and narrower tires. A lighter vehicle presses down on the tires less and reduces road friction.

Tire pressure is the *single largest factor you can immediately control*. When temperatures drop, the pressure in your tires decreases. For every 10 degrees reduction in outside temperature, cold tire pressure is estimated to drop one pound per square inch (PSI).

According to Goodyear Tire Company, "Running a tire 20 percent under inflated – only 5 to 7 pounds per square inch – can increase fuel consumption by 10 percent. That can easily cost motorists two or three miles per gallon. Not only that, but the tire's tread life is reduced by 15 percent".

Here is a good source of additional knowledge for tire care, including advice for tire selection, inflation, rotation and tread wear inspection.

http://www.goodyeartires.com/kyt/

http://www.michelin.com

Mechanical Friction

Rubbing parts are the source of mechanical friction. Any time two moving parts move against one another as in the case of engine crank-shafts, connecting rods, camshafts, valves, bearings, gears, belts, and brakes, there is friction, heat loss, and resulting losses as energy is created in the vehicle's engine combustion chambers and transferred from each cylinder through the engine, drive-train, tires, and to the ground.

Science... or Science Fiction?

Did you know that the science of rubbing parts is called Tribology? There is an entire industry with its own science out there just for rubbing parts! They even make up their own big words. To us, Tribology sounds like the art of caring for Tribles (as seen many years ago on Star Trek)...

There is a tremendous amount of research looking into creating frictionless machines, bearings and lubricants. Modern technology has made some major advances and has produced a number of materials and lubricants that greatly reduce the amount of friction encountered in a vehicle. We are seeing Teflon coatings and synthetic bearing grease, among other products, available today.

Wear and tear is a natural by-product of mechanical friction. Gears wear out, belts crack and break, bearings wear thin and brake pads wear out, all because of friction.

We will show you some alternative methods of applying high tech lubricants in your vehicle and perform other adjustments to greatly reduce friction and enhance fuel efficiency.

Hydraulic Friction

Pressure and viscosity of a fluid being forced through a channel dictate how much hydraulic friction is present, as does the material the channel is made of.

That Sucks!

If you've ever tried to drink a thick milkshake or smoothie through a straw, you know what we're talking about. If the milkshake is too thick, it takes a tremendous personal effort to suck that viscous fluid up through the straw (and maybe get a hernia, too!).

The same principle applies in your vehicle. Engine oil is designed to be a certain thickness, or viscosity under varying conditions of heat and cold. The more viscous the oil, the harder the engine has to work to continually power the oil pump to circulate the oil throughout the engine.

In general, we want to *use the lowest viscosity oil* that will protect and lubricate our engine based on the climate and temperature extremes in which we operate.

The higher the pressure of the system in which the fluid is being circulated, the more energy required to move it. Many fluid systems work under pressure in your vehicle including the oil, braking, power steering, air conditioning and coolant systems.

All of these systems draw energy from the engine to allow them to work. So, the cleaner and more efficient each of these systems is, the lower the overall energy drain on the engine.

Tips and Techniques for Reducing Friction Losses

Solution #1 **Keep your tires inflated properly**

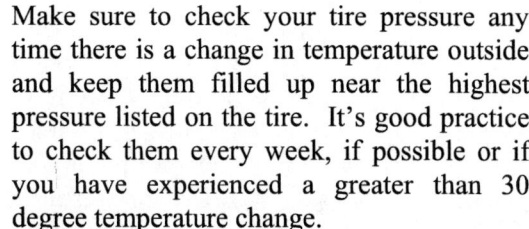

Make sure to check your tire pressure any time there is a change in temperature outside and keep them filled up near the highest pressure listed on the tire. It's good practice to check them every week, if possible or if you have experienced a greater than 30 degree temperature change.

Check for abnormal wear, typically along the center line of the tire if over-inflated and along the edges if under-inflated. Uneven wear or cupping of the tread along one or both edges is an indication of an alignment problem or an issue with your shocks or struts.

This may be one of the most important ways for you to keep your fuel mileage high. Remember, adjusting your tire pressure for *your specific vehicle load* is important for optimum fuel mileage.

How many times have those of you with work trucks loaded the bed with 1,000 to 1,500 pounds of cargo or trailer tongue weight and just driven on, unaware of the consequences. Have you ever noticed your work trucks' tires protesting this increased rolling resistance load and squished like pancakes!

We've all seen this at one time or another. Both safety and efficiency are clearly at risk and are sacrificed every time you see this situation.

Solution #2 **Don't drive with your windows down**

In another section, we will suggest driving without Air Conditioning unless you really need it. There is a trade-off here. If it is too warm to keep the A/C off AND keep your windows up, you are better off running your A/C than you are driving with at highway speeds with your windows down because of the excessive drag created by the open windows.

Solution #3 **Apply synthetic bearing grease to your wheel bearings**

Good quality synthetic bearing grease will greatly reduce the friction around your wheel bearings and will help decrease overall rolling resistance.

Try products from Mobil 1, Red Line and Amsoil and follow the minimum recommended specifications for all lubricants including bearing grease. They are all 25+ years in the business with NO F.T.C. (Federal Trade Commission) complaints levied against them. As in this case and others in this book, the "good ones" seem to be those that have the staying power of tens of years in this market with no complaints. Remember the old saying, "Don't discuss it and let the actions speak for themselves?" These three have done a good job of doing just what they say they are going to do. There are so many others that shout from the highest perch about what they can do and levy outrageous claims. Our experience tells us the quiet performer stands the test of time, not the "town-crier".

Solution #4 **Use synthetic oil in your transmission, differential and transfer case**

Good quality synthetic ATF or appropriate gear oil will dramatically reduce the friction and operating temperature inside your automatic or manual transmission, differential(s) or transfer case.

It could also last up to 50,000 miles, or 3 to 5 times longer than conventional petroleum based oils. Some manufacturers are already placing full-warranty maintenance in new 2006 vehicles at up to 100,000 miles for transmissions. Please review your owner's manual and/or visit your local dealership to get their advice.

If you choose to explore this, talk to your service writer or mechanic. *They are a huge wealth of information.* We have been talking with mechanics like this for over 25 years. *This is where you find out what's really going on. You can get the inside scoop about your vehicle. You find out what works and what doesn't work. You learn about typical problems and the solutions they use to fix them.* Remember, in many cases it is not what you know but who you know.

There are hundreds of millions of dollars spent rebuilding and replacing "worn-out" or under-maintained transmissions each year. Having the correct knowledge from the experts here can go a very long way toward increasing your transmission's efficiency and longevity.

Here are a few transmission and gear lubricant manufacturers who create quality products that meet or exceed all automaker's specifications for transmission and gear box lubricants:

- AMSOIL Long Life Synthetic Gear Lube SAE 75W-90 (change interval up to 500,000 miles!) as well as ATF.

- AMSOIL SEVERE GEAR™ Synthetic EP Gear Lube SAE 75W-90 (change interval up to 100,000 miles) for heavy duty or severe operations as well as ATF.

- Red Line 75W90 High Performance Gear oil as well as ATF.

Solution #5 Make sure your wheels are aligned to reduce rolling resistance

Maintaining your vehicle's alignment is paramount to the operational efficiency and safety of your vehicle. How many times have you (or your significant other) bumped a curb or hit a pretty significant bump? Did you know one small uneventful bump from a curb can significantly mis-align your car?

Most people are un-aware that if you have a "cantered" steering wheel while driving straight down the road, you are probably in need of an alignment. We see drivers most every day "crabbing" down the road. This is a big sign you need to have your alignment checked and possibly tuned.

The next time you are going down the highway take notice of the position of your steering wheel. It should be in alignment and as level and straight as the level road you are traveling down. If your steering wheel is not straight then imagine it straight. DO NOT DO THIS WHILE DRIVING as it is unsafe - But imagine if you were to center the wheel from its canted position, then how far from straight would the car deviate?

Constantly traveling down the road "sideways" (figure of speech, but you get the point) is horribly inefficient and a big cause of premature wearing of your tires. This can be a double whammy in that you pay more out of your pocket for each inch you travel because of poor gas mileage caused by an out of alignment vehicle AND having to buy new tires more often. Sounds a bit simplistic but this is a commonly overlooked mileage eater.

Solution #6 **Maintain your brakes and make sure your emergency brake is disengaged**

There are two parts to the brake equation - your standard brakes and your emergency brakes. Both are absolute critical systems and the highest priority should be given to their maintenance and effectiveness. But what do brakes have to do with getting better gas mileage? The answer is simple - worn and improperly maintained brakes can cause excess drag and may become a driving hazard.

The brake pad contact points can fall out of alignment, ever so slightly, and create unwanted friction between the pads and the rotors. In most cases, you will hear a slight chatter or chirp telling you there is a problem. You may also feel this as slight vibration or chatter with your foot as you apply pressure on the brakes.

Additionally, today's newer cars, with 4-wheel disc brakes, have a second-set of brake shoes (emergency brakes), used to keep the car from rolling when parked and to stop the car in the event the primary system has failed.

All too many times drivers leave the parking brake partially engaged and do not figure this out until they see smoke. If the emergency brakes get so hot they begin to fuse the shoe to the friction material it is in contact with, you may feel a distinct chatter or vibration in the car.

Solution #7 **Make your vehicle more aerodynamic**

Anything attached to the outside of your vehicle will add extra wind resistance. A car-top carrier, a bike rack on the back, even your luggage rack will add wind resistance. If your luggage rack is removable, take it off when you don't need it.

Here's a good one for truck owners. It is common thinking that if you take the tailgate off, you will decrease the wind resistance and get better gas mileage. That is the reason you see so many trucks with a cargo net across the back of the truck bed instead of the tailgate.

We learned from an an aerospace engineer the truth about this tailgate mystery. It seems that with the tailgate on, there is a "bubble" of air that forms in the truck bed. The air passing over the top of the truck also passes over that bubble of air, as if it were a part of the vehicle, making a fairly smooth path for the passing air. We already learned the smoother the flow of air, the less wind resistance there is.

Now, if you remove the tailgate, the bubble of air collapses and the passing air swirls and becomes turbulent in and around the truck bed, causing more wind resistance.

We were pretty amazed at this explanation, but it makes a lot of sense. Bottom line – keep your tailgates in fully upright and "locked" position, truck owners!

A lot of long-haul truckers are learning the value of streamlining their trucks, and the same lessons apply to regular cars, trucks and SUV's. Race car drivers know this too. If you can add an air dam, spoilers, or fairings around the bottom of the vehicle to minimize the amount of air that can pass underneath, the vehicle will be more aerodynamic and the engine won't need to work so hard to move it. Obviously, if you truly go off-road in your SUV or truck then this may not be the option for you.

Reducing Mechanical Load Losses

Mechanical Load Loss due to Engine Subsystems

Subsystems powered by the engine are vital for the operation of the engine itself, as well as for systems in the vehicle for passenger comfort and safety such as heating and cooling, electrical, steering, braking, and other functions.

Subsystems required for the engine itself to operate include the ignition, fuel, coolant, and air intake systems. All of these systems burden the engine in order to run. But it is a self serving effort for the engine, because without them, it wouldn't be able to run.

The engine subsystems absorb a considerable amount of the engine's output power. That is why you hear in the conventional fuel economy advice that you should minimize use of the air conditioning system, which is in fact good advice.

The engine's internal parts also require energy to push, pull, rotate and spin. These internal components include primarily the pistons, crankshaft, camshaft and valves. Friction, as discussed in the previous section, also plays a key role in the total amount of engine output power being used.

Following is a list and a brief description of each of the primary subsystems and how they burden the engine:

Air conditioning. The air conditioning compressor is driven by a belt connected to a pulley on the engine. When the A/C is turned on, it takes a great deal of power to turn the compressor pump to cool the air. Electric fan motors inside the vehicle also draw electricity from the belt driven alternator and cause even more load on the engine.

Heating the car is not as stressful to the engine, as the heat is naturally drawn from the hot fluid circulating through the engine and keeping it cool. Fan motors inside the vehicle, however, do draw power from the alternator and load the engine that way.

Belts and pulleys. Most every vehicle engine has a series of external and accessory drive pulleys being driven by the internal mechanism of the engine. These pulleys drive a variety of subsystems such as electric, a/c, steering, and braking via a belt. As more and more subsystems are being driven by these belts and pulleys, the heavier the load on the engine.

Some "hot-rodders" and others advocate using aftermarket "under-drive pulleys" to decrease sub-system load and increase the overall output of the engine. This increased output comes at the price of sacrificing overall balance of the engine, namely the cooling and electrical systems.

These under-drive pulleys "slow" the rotational speed of the pulley compared to standard pulleys. For the cooling system, this means a slower flow of coolant through the engine. This means a heavy load on the engine for any length of time will probably tax the cooling system so even with a fully open thermostat the engine could overheat. Some people put lower temperature thermostats in their engines to try to compensate for the decreased coolant flow. This doesn't always work. Remember, if it were a truly viable engineering solution then the new vehicle manufacturers would put this in their new SUV's and trucks.

There is nothing quite as terrifying as watching the temperature gauge, in your fully loaded Suburban on a 100 degree day, jump from 150 degrees to 275 degrees because you slowed down in traffic and don't have enough coolant flowing through the engine to keep it cool. That exposes your engine to dramatic and possibly catastrophic temperature variances.

Alternators turning at a slower speed will likely fail sooner because the voltage regulator (typically inside the alternator) has to operate far outside its normal operating tolerances, shortening the life of the alternator. For vehicles with a heavy electrical load, it is feasible to entirely overpower the electrical system in this manner. This one seemingly benign under-drive pulley change puts two other significant management systems at operational risk or even failure.

Fun Flashback Note:

Lining up to race my '57 Chevy against a friend's 67' Chevelle back in high school, I was curious when he got under the hood and unhooked all the belts from his engine. I now understand why he did that, to allow more of the engine's output power to go directly to the wheels and not get drained off. I'm sure that's why he beat me (barely) ☺. Wonder if I could track him down now for a rematch...?

Electrical. Vehicles nowadays are equipped with every modern convenience it seems, especially SUV's. What would we do without the dual air conditioning systems, surround sound stereo, DVD player, television tuner, satellite communications (OnStar), satellite radio, 110 volt inverters with electrical outlets, multiple power ports for laptop chargers, cell phones and other high tech gadgetry.

In some cars there are the massive stereo systems with subwoofers, amplifiers and all the other gadgets creating the BOOM BOOM sound that rattles all cars within a half mile radius and shakes the earth. You know what I'm talking about.

Where does the energy come from to power all this stuff? The engine of course! A little device called an alternator is the electrical power plant for your vehicle. It is driven by a belt from the engine and generates a significant amount of electricity to power the growing electrical needs of today's vehicles.

In addition to the components listed above, the alternator also powers all the electricity for everything else in the vehicle including the lights, the computer systems (in newer vehicles), dashboard instruments, windshield wipers, fuel injectors and the ignition system.

This is truly a workhorse subsystem, and as you can begin to imagine, draws a tremendous amount of the engine's output energy to convert to electricity to support its load.

Water pump / coolant. Engines create a substantial amount of heat caused by the mini explosions occurring inside your cylinders repeatedly. If left unchecked, that heat would begin to warp and melt the various parts of your engine and render it useless after only a few minutes of operation.

Thank goodness, then, we have a way of keeping the engine cool. The coolant system circulates fluid (that green or red liquid) through a hollow shell around the engine where it picks up heat. The fluid then passes through the radiator where the outside rushing air passes through and cools it again. The coolant is pumped back into the engine again to "pick up" more heat for disposal. A water pump, driven by the engine – yet another source of energy loss – is responsible for moving the coolant around and around through the engine and radiator.

Oil system. Friction reduction and internal heat dispersion is a job for your engine oil. All those fast moving parts need constant lubrication to minimize friction and to keep them cool (cool being a relative term – the inside of the engine is many hundreds of degrees F.).

Without the oil doing its job, the metal parts would quickly heat up and expand from the combined effects of friction and engine operating temperatures. The expanded parts now have even more friction, causing more heat and more expansion. Very soon, the engine can no longer overcome the incredible frictional force being generated and "seizes up".

This is a bad thing. (Something that just happened to my 17 year old son!!)

Safety Tip

Whenever you see a low oil pressure reading, whether on a gauge or a so-called "idiot light", pull the car off the road at the first safe opportunity, shut the engine off and call for professional help. Don't delay on this. My son thought he could make it another couple of miles to the house when his light came on. Didn't make it. Now I'm replacing the engine. Don't let this happen to you!

The oil pump is located at the bottom of the engine, pumping oil from where it settles in the oil pan, up and throughout the engine. It creates a pressurized system inside the engine to circulate the oil efficiently. Like all the other subsystems, it puts a load on the engine in order to perform its vital function.

Fuel system. Delivery of clean, high quality fuel to the engine is of utmost importance. We will discuss this system more in a little while, but for now, we will just mention that the delivery of the fuel requires a pump to move the fuel from the fuel tank to the engine via an in-line fuel filter.

Then the fuel is squirted through a fuel injector (one injector for each cylinder) to mix with the air and create an explosive mixture that will be ignited and explode. The fuel injectors (on newer trucks and SUV's) and the pump are electrically operated and draw energy from the alternator for their operation.

Ignition system. Up to 100,000 volts of electricity passes through each spark plug creating a hot enough spark to ignite a compacted mixture of atomized fuel and air.

The ignition system performs a sophisticated dance to send the right jolt of electricity to the right spark plug at just the right moment to light the fuel/air mixture. The mixture explodes just in time to drive the piston downward, thus creating power in the engine.

This can happen thousands of times per minute in your vehicle's engine. Needless to say, this can be quite a draw on the vehicle's electrical system.

What's interesting is that the engine is creating the energy to drive the electrical system which is providing the energy to drive the ignition system, which is powering the engine! A self-sustaining operation, fueled by gasoline – pretty cool, huh?

Power brakes. In the old days, we had to have a strong leg to push on the brake pedal and slow the car down. These days with the advent of disk brakes, the brakes have a power assist function driven by a vacuum booster which draws vacuum from the engine.

The power assist function reduces the amount of force our legs must apply on the brake pedal and applies a greater force on the hydraulic wheel brakes so we don't need to! The vacuum drawn is a small but noticeable energy loss from the engine.

Power steering. We were also much more buff in the old days with bulging biceps and forearms of steel forged by the exercise of turning our steering wheels without power steering. OK, so maybe not so buff, but we did need to work pretty hard sometimes to turn those things! Power assisted steering now is also hydraulically driven through the power steering pump which is driven by a belt from the engine.

Engine moving parts. Moving parts of the engine, while not technically considered a subsystem of the engine, impart a load on the engine nonetheless, so we will address this load briefly.

We'll discuss more how the engine works in the next section but for now; understand there is a symphony going on in there. Pistons, valves, crankshaft, camshaft and other moving parts are big, heavy components operating under high pressure, heavy forces and extreme heat.

If you've ever tried to crank an engine by hand, even for half a turn, you'd know what we are talking about. It takes a strong person to do that.

So, consider that an engine has to create enough energy to turn itself, power all the subsystems we've just discussed and then provide power to the drivetrain, which will transfer that power to the wheels.

Mechanical Load Loss due to Drivetrain

The drivetrain is the collection of everything between your engine and your tires that converts energy from the engine to forward momentum in your vehicle. It generally includes the torque converter, transmission, drive shaft(s), gear boxes and wheel assemblies.

Following is a quick discussion of the main components of the drivetrain. For a more thorough discussion about how a drivetrain works, visit *www.howstuffworks.com*.

> **Why should you care how your car works?**
>
> Many people want to know why something is true and how it works before they believe it to be true. Others are simply curious about what is going on with all the spinning, whirring, rumbling and belching under the hood. Finally, there are those who just don't give a darn about how or why these gas mileage tips work – they just accept that they do.
>
> We provide the "how it works" information in this book to satisfy anyone who wants to know. It is also a proven fact that you are more likely to behave in a certain way if you understand the consequences of not behaving in the desired way. For example, if you understand that it takes eight times more horsepower for you to drive at twice the speed, you get a sense for how much gas that must use and decide to drive a little slower.

Transmission / Transaxle. Converting energy output from the engine to the driveshaft, gear boxes and wheels is accomplished by this mysterious device we call a transmission. Some transmissions are automatic and others are manual, but all perform the same basic function.

The transmission allows the engine to operate within a narrow range of speeds while allowing the car to operate in a wide range of speeds. This allows the engine to operate at the speed where it delivers the most torque and horsepower, where it is most efficient as specified by the new vehicle manufacturer.

Manual and automatic transmissions are very different on the inside but provide the same result. The function they perform is vital, but they do charge a tax for their services, as they exert an additional load on the engine as they convert engine output to drivetrain output.

Transaxles are found on either front or rear wheel drive vehicles where the transmission and the drive axle are combined in the same unit. The transmission portion of the transaxle operates in exactly the same manner as other transmissions. In the case of a transaxle, a drive shaft is not typically required.

Drive Shaft. Rear-wheel drive vehicles need a mechanism to transfer the output of the transmission from the front of a vehicle to the rear wheels. This mechanism is a drive shaft. It begins at the rear of the transmission and traverses the length of the vehicle where it connects to the differential on the rear axle.

The drive shaft is a heavy and precisely balanced metal shaft connected by universal joints on either end for flexibility as the parts of the drivetrain move, twist and bump around.

In the case of four wheel drive vehicles, there is a transfer case just behind the transmission interconnecting a front drive shaft for the front wheels to the rear drive shaft. The front drive shaft attaches to a differential on the front wheel axle. Four wheel drive systems inherently add weight and complexity to any SUV or truck. By their very nature they reduce on-road driveline efficiency for the added security and stability of four wheel drive in slippery or off-road conditions

Differential / Transfer Case. Spinning tires create a problem for the vehicle, so a clever device has been created to manage traction on the drive axle where the slipping occurs. Let's say you are beginning to accelerate and your primary drive wheel begins to slip in the snow.

The differential senses the slippage and engages the other wheel on the same axle for added traction. There are different applications of that basic scenario, and some cars do not have it, but many of the larger gas guzzlers we are discussing here fall into that scenario.

Four wheel drive vehicles have a differential on both front and rear axles, connected by drive shafts to the transfer case, which locks the two drive shafts together when the vehicle is placed into four wheel drive mode.

The primary roles of the differential are to transfer the engine output energy to the wheels, to provide the final round of reduction in gearing, and to allow the two wheels it manages to spin at different speeds (such as when cornering or regaining traction).

There is a significant mechanical load and friction loss during the transfer of energy from the drive shaft to the wheel assemblies via the differential(s) and transfer case.

Wheel Assemblies. Allowing the vehicle to roll down the road is the job of the wheels on your vehicle. This is where the rubber meets the road, so to speak.

The wheel assemblies include the tires and wheels, the brake assemblies, and the steering controls. Together these assemblies are the controlling mechanism for moving, turning and stopping your vehicle.

We've already discussed the rolling resistance of the tires on the road, which are a significant drain of energy for the vehicle.

During acceleration or constant speed driving the engine has to exert enough energy to keep the wheels rolling forward at the desired speed.

When slowing or stopping, the brakes apply pressure to the brake pads or drums to convert forward momentum (kinetic energy) to heat.

Unfortunately, this is a *very inefficient way to manage energy*, since every time we use our brakes; we are *converting hard earned momentum into heat that we can no longer use*. We will address this at length in the section on Driving Technique.

This is a very important aspect of regenerative braking, which some new hybrid and advanced technology vehicles are using to recapture the energy otherwise lost to heat while braking.

Mechanical Load Loss due to Weight/Mass

This is a common sense topic. *If it weighs more, you must work harder to move it.* Same thing with your gas guzzler.

If your four wheel drive F-350 has the heavy duty brush guard and rear bumper, heavy duty suspension, winch, racks, tool boxes, big tires and wheels, AND full of tools or cargo, AND your passengers, we'd guess you're sporting an extra couple of thousand pounds of driving weight.

Yes – that's a TON!

Now, most vehicles don't (can't) support that kind of payload. But, we'd be willing to bet you can find lots of spare baggage in your trunk or hatchback.

We completely understand some people have no choice about carrying extra cargo. Just be aware that if you have the opportunity to lighten the load, you will reduce the load on your engine and get better gas mileage.

Tips and Techniques for Reducing Mechanical Load Losses

Solution #8   Remove all unnecessary baggage and payload from your vehicle

Most of us carry around a lot of junk in the back of our Truck or SUV. I (Ron) just went out to look at mine and found probably 30 pounds of soccer gear, tile samples, cleaning supplies, folding chairs, and miscellaneous clothes (my kids use the car as a changing room all the time). I also found 80 pounds worth of tools and totally random items in my truck tool box. Minimizing the junk reduces the amount of weight the vehicle has to move carry. Over time, even small amounts of inefficiency add up.

The Department of Energy estimates you will gain 1-2% improvement in fuel economy for every 100 pounds you shed. That's the equivalent of paying $.03 to $.06 less per gallon at the pump!

Solution #9 Keep your air conditioning turned off when it is not hot

As mentioned in the previous chapter, keeping the air conditioning off when possible and the windows rolled up yield the best fuel economy results. Rolling down the windows at highway speeds generates additional turbulence and drag. If it's too warm to keep the windows up and use your vents in the car for cooling, you're better off running your air conditioning.

In today's climate controlled vehicles, the air conditioning is typically left on at all times, even when it is cold outside. Meaning, the compressor is running constantly, pulling energy from the engine, and "conditioning" the hot air being blown into the car or truck. This is, in most cases, not necessary. You should be able to keep your heater running without the air conditioner being on. This simple little tip will save you lots of money in mild and cold weather.

Excessive use of the defroster, according to the EPA, is just like running the air conditioning and can consume extra energy from your engine to do so. Just remember to use it only when necessary and then turn it off.

In warmer climates where air conditioning is prevalent, a reduction in air conditioning usage can be realized by tinting your windows to the darkest shade that is legal. Besides lowering glare in sun-drenched areas, the window tint also reduces interior ambient heat. This allows you to run the air conditioning less and allows the A/C compressor to cycle or disengage more often. The lightened A/C compressor load on the engine will allow better mileage.

Solution #10 Keep unnecessary electrical accessories turned off

Keep all your unused electrical accessories turned off. If you are not watching the video gear in the back seat, or if you don't need the electric cooler right now, turn it off. Keep in mind your high power stereo speakers, subwoofers and amplifiers put a heavy load on your vehicle electrical system and the engine.

Solution #11 Use high quality synthetic oil in your engine

Use high quality synthetic oil in your engine. It may seem more expensive at first, but when you realize you need only change your oil every 10,000 miles, you'll be a believer!

There are advertisements, even today, that companies such as Mobil 1 openly offer synthetic oils having a maintenance interval (on the container) of 15,000 miles! Other reputable synthetic oil manufacturers stand behind even longer change intervals!

Here are some good quality brands to help you decide which oil brand and viscosity is best for you:

Red Line Synthetic Oil

Castrol Synthetic Oil

Amsoil Synthetic Oil

Mobil 1 Synthetic Oil

We recommend you change your synthetic oil at intervals no longer than 10,000 miles, even if longer intervals are recommended by the manufacturer. Oils tend to pick up contaminants from the combusting air and fuel in the engine. As time goes on, the build up of contaminants causes the oil to become acidic and cause bearing damage.

So, even though synthetic oils could theoretically maintain good lubricating properties for 15,000 or even 25,000 miles, we feel it's best to change your oil at least every 10,000 miles.

Change the Oil Filter half way through!

Follow the synthetic oil manufacturer's recommendations and change your oil filter half way through (for example, at the 5,000 mile mark). The oil is still picking up contaminants and needs a new filter so that it can do its job for the full 10,000 miles.

Additionally, having your automoti[cut]
tested at various intervals likens to [cut]
your blood tested as humans. We [cut]
recommend having your oil teste[cut]
reputable oil testing laboratory. So[cut]
labs performing these tests are:

http://www.oaitesting.com

www.analystsinc.com

Caution about Oil Additives

Please be very, very cautious about using oil add[cut]
engine. There continues to be a great deal o[cut]
around whether they are really helpful, or wh[cut]
harmful. There are also lots of scams and rip-o[cut]
to get your hard-earned money.

We DO NOT RECOMMEND that you use oi[cut]
are using good quality oil in your vehicle.

The ONLY time there MIGHT be a goo[cut]
additive is if you have an old clunker that [cut]
keep) running. Under these circumstanc[cut]
have anything to lose, we don't see the h[cut]
Make sure to do your homework first, a[cut]
compatible with the oil you are using.

In general, you [cut]
viscosity oil recor[cut]
Good 5W-30 wei[cut]
choice for most [cut]
the synthetic oil [cut]
viscosity oil, [cut]
recommend thi[cut]

Solution #12 Consider installing an electric radiator cooling fan

An electric radiator cooling fan is more efficient than a belt driven fan, turning on and off as required. The mechanically driven belt fan is always exerting a load on the engine. Conversion kits are available for most vehicles. Why don't the automakers put electric fans on all vehicles,? Because they are more expensive. We believe more and more automakers will begin adding electric fans to new vehicles to meet the new mileage demands of the buying public.

Solution #13 Keep your fuel filter clean

A properly maintained fuel system with clean, filtered fuel and clean fuel injectors ensures a baseline of dependable fuel flow to your engine. We'll discuss keeping fuel injectors clean in Chapter 4, Optimizing Engine *Efficiency*.

To minimize any additional drain on the electrical system, we want the fuel pump to operate as smoothly and easily as possible. Changing out the fuel filter frequently is the best way to ensure clean fuel and the least burden on the fuel pump. A dirty filter requires the pump to work extra hard, drawing additional energy from the alternator, which in turn, puts more of a load on the engine. Over time, eliminating this small amount of extra load can add up to some pretty big savings.

The experience of replacing a failed fuel pump just happened to our (Kenny's) test mule when purchased for the road testing and validation of this book. The cost to replace the fuel pump on a 1999 Suburban was close to $900.00! Don't let this happen to you!

Solution #14 **Ensure your power steering fluid and belts are properly maintained**

If you have ever heard the groan and whine of a power steering compressor low on fluid then you know what we mean. Low fluid levels can cause early wear and failure of your power steering system. Having your belts properly adjusted is will allow your engine to provide energy to your power steering system without being overburdened. We recommend a good synthetic power steering fluid like Mobil, Amsoil, or Red Line.

Solution #15 **Make sure you are using the proper coolant in your vehicle.**

Do not use water only in your cooling system. Water alone does not have the capacity to keep today's engines running cool. A cool running engine is more efficient and lasts longer.

Today there are anti-freeze brands that are good for as much as 100,000 miles! These coolants have special rust inhibitors as well as cooling stabilizers increasing the boiling point to well above that of water. Look in your vehicle owner's manual to determine the right type of coolant to use.

Reducing Heat and Noise Losses

Losses due to Heat and Noise

According to the Rocky Mountain Institute (RMI), respected advocates of conserving natural resources, internal combustion engines only capture 15 to 20% of the potential energy contained in gasoline. The other 80 to 85% is lost as heat and vibrational noise.

Think about this for a moment... over 80% of the potential energy in each gallon of gasoline is lost and never gets used by the engine at all! The rest of the energy is released as unburned hydrocarbons in the exhaust system, or lost as heat. All that heat emitted by the engine is largely wasted.

The only time engine generated heat is put to any use at all is when the vehicle heater is turned on to capture a small fraction of that heat for passenger comfort. Could you imagine what it might be like if we could capture the other 80% or so of that lost energy? Mind-boggling, huh?

Heat itself is not a *primary cause* of energy loss in this situation. It is, in fact, the *result* of energy loss. In the big scheme of things, you can not destroy energy; you can only convert it into another form.

Heat is a common by-product of an inefficient energy conversion process. In the example of the internal combustion engine, the energy in gasoline is converted to mechanical energy via a chemical reaction (combustion).

Roughly 20% of the energy goes to power the engine and ultimately move the vehicle forward. The other 80% is converted to heat or lost as unburned fuel.

Now, here is where heat becomes an indirect source of energy loss. If the engine gets too hot, or in other words, if our cooling system isn't efficient enough to pull the heat away from the engine, the excess heat leads to other energy draining effects. The combustion process is less efficient as a result, and the engine components expand from the heat, incur additional friction losses, and wear out much sooner than they should.

Everywhere there is friction, there is energy loss. The less friction, the less energy lost, and the less energy is converted to heat. Therefore, any attempts on our part to reduce friction will reduce energy loss, and will reduce the build-up of excess heat.

An Apple a Day....

Remember the saying about how to keep the doctor away? Well, the same thing goes with your car or truck. Keeping it well maintained with a good tune-up, good oil (with frequent changes), and with a good coolant will keep your vehicle running smoother and cooler and will avoid much of the energy loss that occurs with poorly maintained engines.

Tips and Techniques for Reducing Heat and Noise Losses

Solution #16   Clean your engine at least once a year

Steam clean or de-gunk your engine at least once a year or more often if needed to keep it clean. Buildup of grease, oil, dirt and road grime captures and holds engine heat, causing your engine to run warmer than necessary. Keep it clean and keep it cool!

Energy Losses – The BIG Picture

So, let's summarize the energy equation here. An internal combustion engine captures only 20% of the potential energy of a gallon of gasoline. Of the 20% of energy converted to mechanical energy, over half is lost to friction, mechanical load, heat and weight/mass losses through the engine subsystems and the drivetrain. The actual amount depends on your vehicle, the driving environment and your driving habits.

As a result, we need *bigger and more powerful engines* to operate our gas guzzlers, since only a small fraction of the energy we use ever makes it to the point where the rubber meets the road.

Kenny Joines & Ron Hollenbeck

Chapter 4: Optimizing Engine Efficiency

Internal Combustion Basics and Optimum Tuning

In the last chapter, we discussed minimizing the losses occurring from friction, mechanical load and heat. In this section, we will discuss making your power plant as efficient as possible.

Combustion of the fuel/air mixture occurs as the piston is reaching the top of its travel. This explosion of the combined fuel and air happens just in time to drive the piston down with thunderous force, causing the heavy metal crankshaft below to rotate.

Multiple pistons working together in just the same manner in their own cylinders keep the crankshaft turning smoothly. This is the source of power propelling your vehicle forward.

Timing is critical inside the engine as fuel and air must be delivered in just the right quantity, the right consistency, and at the right time.

A spark is introduced at just the right moment, even as the piston is closing in on the fuel/air mixture, compressing it so that it is even more volatile when the spark ignites it.

Please visit *http://auto.howstuffworks.com/engine.htm* on the web to get more detailed information about how an engine works. You will also find some very cool animations showing just how the pistons move up and down in the four stroke cycle and the related timing of all the other events.

In spite of the efficiencies and technological know-how carmakers have built into the internal combustion engine, even after more than 100 years of experience, the internal combustion process is inherently inefficient. At best, we extract 20% of the available energy from a gallon of gasoline. The rest is lost as heat and vibrational noise.

We have become much more efficient at harnessing the 20% of usable energy for use in powering the vehicle and all its subsystems. Still, there are inefficiencies we will point out to you so you can take steps to improve how well your engine operates, and improve your fuel economy in the process.

Why Automakers Don't Engineer and Tune their Vehicles for Optimum Mileage

Many of the inefficiencies we discuss are known by the automakers, but for a variety of reasons these inefficiencies are not optimized at the factory.

As we mentioned briefly before, automakers build and tune their vehicles to the lowest common denominator. They study typical driving habits, locations, environment, elevation and other factors and come up with an "average" factory tune as the best compromise of performance, mileage and emissions for each vehicle.

Automakers must also comply with much stricter emissions and noise restrictions than makers of aftermarket products do. While they do have the budgets to perform extensive research and develop highly specialized components for your vehicle, aftermarket product makers don't have as many rules.

Aftermarket manufacturers have a more narrow focus and can target their budgets for developing specialty products. Automakers have a much broader area of interest which tends to dilute their focus and their budgets.

If automakers put that much time and money into doing research and development, the price we pay for our vehicles would probably be much higher.

That is why there are a number of things we can do ourselves, and a number of very high quality aftermarket products to help us achieve the "perfect" tune for us.

> **"Right-Foot-Planted" Syndrome – The Bain of Better Fuel Economy**
>
> Ever notice how fast most people take off from a stop at a red light? Odds are you haven't because you are right in there with the rest of the pack.
>
> Most people in the U.S. do what we call a high-acceleration take off from a stop. The good thing about this is – well, we're not sure what the good thing about this is, except maybe to prevent getting run over by the guy behind you.
>
> The bad thing about this syndrome is that it KILLS your gas mileage. It's also contagious and habit-forming. And, it's even more addicting when you've gone to the trouble of making your car or truck more efficient and thus, it has more performance. It's FUN to feel that added power.

"Right-Foot-Planted" Syndrome – continued

Unfortunately, any gains in fuel economy from following the tips and techniques in this book will be INSTANTLY negated and you may actually experience POORER gas mileage when you "plant" your right foot accelerator to the floor after each stop. If you keep these same "right-foot-planted" habits, yes, you may see greater acceleration speed but an even greater bill at the gas pump! Even an extra half inch of unnecessary depression on your gas pedal can mean the difference between gaining and losing that extra 2-3 miles per gallon in your gas guzzler. And then you'll call us and want your money back because you're not getting better fuel economy!

Look at your owner's manual for the prescribed "shift-points" of your vehicle's transmission. Then, go out and try to "drive" your vehicle in the manner where you owner's manual says the earliest "shift-point" occurs. That is NOT to say to floor it to make it shift faster! Just the opposite behavior needs to occur.

The harder you accelerate, the faster your engine rev's before each shift point. If you accelerate gently, the transmission will shift at a lower RPM. Remember, higher RPMs in these vehicles means higher gas consumption!

Try to drive the "smoothest" you can to see when the transmission (Automatic) will shift into the next gear at the slowest physical speed. If you try to remember to drive this way all the time, you will be keeping more money in your pocket and putting less in the oil company's pocket!

Six Components Affecting Engine Efficiency

There are six major components affecting engine efficiency over which we have control.

- Air delivery
- Fuel quality and delivery
- Fuel/Air combustion, ignition and timing
- Exhaust
- Lubrication and cooling
- Mechanical health and tuning (bearings, rings, seals, cylinder wear, valves)

We will discuss each of these areas briefly and then introduce recommended solutions for improving efficiency in each area for your engine.

Optimizing Air Delivery

Huge amounts of air are required by an internal combustion engine for it to perform well. The incoming air is mixed with atomized fuel and then compressed in the cylinder to create an explosive mixture.

High air flow is critical for efficiency and performance of your vehicle. If the engine is getting sufficient air flow throughout its full operating range then the Electronic Control Module (ECM), which in modern cars is the computer brain for the vehicle, will be more likely to stay within its standard table of operating parameters, which is to say it will operate most efficiently.

The air being ingested should be clean and cool. To clean the air there is an air filter, typically made of multiple layers of very thin paper or oil coated cotton gauze material that trap dust, dirt and other particles, yet allow the air to pass through. If the air were not cleaned first, the particles of dirt would quickly damage the engine.

Most vehicles come from the manufacturer with the standard paper filters that do a sufficient job in filtering out the dirt, but they also restrict air flow.

Higher end air intake systems use filters made of multiple layers of woven cotton gauze coated with oil that traps impurities in the air, but allowing a much higher volume of air to pass through.

While paper filters are disposable – once they get clogged up with dirt they must be thrown out – the cotton filters are washable and will outlast your vehicle.

We will not delve too far into the ECM here. Just know that if the engine is getting as much air flow as it needs throughout its operating range then the ECM will keep the air fuel mixture and ignition timing at the optimal point for performance.

The internal combustion engine also prefers cool air, and so you will see automakers go to some length to pull cooler outside air directly into the engine and to avoid pulling in heated air from directly around the engine.

In general, to achieve the best aspiration, it is best for air to travel in a straight line and not travel through multiple bends and turns through an air inlet before arriving at the engine.

The more turbulent the air, the less efficient the fuel/air mixture and resulting combustion. The following solutions will help you ensure enough straight-through, cool air is channeled to your engine.

Tips and Techniques for Optimizing Air Delivery

Solution #17   **Replace your factory paper air filter with a High Flow Replacement Air Filter**

Many aftermarket air filters and devices are designed to flow more air than stock air filters. Paper air filters, which are stock issue in most vehicles, are inferior to oiled cotton gauze filters, provided by the higher quality aftermarket filter manufacturers.

There are a variety of high quality air filters that are exact replacements for most vehicles. High quality replacement air filters are washable and re-usable and will last the life of your vehicle. Replacement air filters from the top manufacturers are emissions legal in all 50 states and they will not void your affect your warranty.

The premise is quite simple, flow as much air volume into the engine through the stock intake system as possible, without impeding the flow with the air filter. At the same time, the air filter must be effective in capturing dirt, dust and other impurities in the air that can damage the engine.

The economics of replacing your stock air filter with a high flow filter are obvious:

- You must replace your stock paper filter sometime soon, anyway
- You must keep replacing your stock paper filter every 15,000 miles or 12 months in most vehicles today
- One replacement high flow may cost as much as 2-3 times more than a paper filter, but you never need to replace it again
- The additional cost of your new high flow air filter will be recovered by increased fuel economy, probably within a matter of months

Here are three of the top high-flow air filter manufacturers:

www.knfilters.com

www.airaid.com

www.afefilters.com

Solution #18   Install an Air Vortex Device

This cool little device, costing less than $75 at the time of this writing, creates a vortex effect with the stock intake air, creating a swirling, fast-burn effect in the combustion chamber. As a result, finer particles of fuel burn hotter and more efficiently. This device is claimed to deliver, on average, an increase of 1-2 mpg and a corresponding increase in horsepower. Results as high as 40% improvement in gas mileage have been claimed by the manufacturers of these devices.

We do not believe (and some of our own preliminary testing seems to support it) these devices are effective in newer vehicles that already incorporate a vortex effect in the air intake. We think the best bet for using this device is in older vehicles only - anything older than the 2000 model year should see an improvement with this device. But, as always, test your results with your vehicle and your situation. These devices don't work for everyone.

Installation is simple and takes 10 minutes to do yourself. Here are a couple of links:

www.turbonator.com

www.tornadoair.com

> **What tangled webs...**
>
> Yes, we know that this air vortex device could be considered a "gadget". And, yes, we know that there are some nay-sayers out there that object to the results being claimed. However, we have personally realized a 1-2 mpg gain over the last two to three years with this device on our older vehicles.
>
> Look at it this way. If you spend $2,500 or more per year on gas, and you get only a 1% improvement while using this device, it will have paid for itself in less than 2 years.
>
> We'll bet that you get better results than that IF you can keep your dang foot off the gas!!

Solution #19   Install a Throttle Body or Manifold Spacer

  Installing a throttle body spacer allows the incoming air to be channeled and "spun" similar to the way an air vortex device spins the air, allowing a more efficient combustion process. Throttle body spacers are available for older carbureted engines as well as for many newer model cars, trucks and SUV's. This addition costs around $100 and an hour or so of your time.

Many Newer Cars Have Incorporated Throttle Body Spacing Concepts

As with the exhaust systems of some of the newer and higher-end vehicles, the engineering has been done to augment this already. Cases of this would be in the 2001 or newer GM Vortec line of engines and in Ford's line of engines corresponding with the same time frame. Additionally, most engines from the Pacific Rim within this time frame have had this increased capacity designed into them already.

This type of change would be for older vehicles that used intake systems using throttle-body injection and carburetion. Tuned-Port injection engines may not see an increase in performance.

Follow the links below to see if spacers are provided for your make and model of vehicle.

The exciting thing about the efficiency improvements resulting from such a spacer are that noticeable improvements are reported at the low and mid range engine speeds, in the range where most people drive every day. Some high end improvement devices only add efficiency at high engine speeds, which only help those looking for a high performance solution and not those who are out for better fuel economy.

Here are a couple of manufacturers of throttle body spacers:

www.airaid.com

www.cranecams.com

Solution #20   Clean and Polish Your Mass Airflow Sensor (MAS)

Your mass air flow sensor detects the volume of air flowing into the engine. The flow of air is slowed down two ways. First, the inside of the valve and air inlet pipe gets all gunked up (technical term) with dirt, grime and other deposits. A dirty inlet pipe and valve can significantly limit the amount of air the engine is receiving.

Secondly, the inside surface of the MAS is made of rough, unpolished metal. The rough surface also creates turbulence and limits the amount of air that can enter the engine. By smoothing and polishing that rough surface, the air flow can be increased significantly enough to yield perhaps another mile per gallon.

Many mechanics offer the cleaning and polishing of the MAS as a service, but it is relatively simple for a shade-tree mechanic to do himself.

Solution #21   Install an Air Intake System

This is a big-ticket item, costing anywhere from $150 to $400 or more, but one that will more than pay for itself if you intend to keep you vehicle more than two more years.

Aftermarket air intake systems allow a larger volume of air with a lower intake restriction. They smooth and straighten the airflow while providing shielding from engine heat and allowing intake of cooler air. We've already discussed aftermarket air intake systems having a reusable lifetime filter that can be cleaned and are far superior to the cheap disposable paper filters.

Mileage and performance gains can be as high as 15-20% with these systems. Here are some high quality Air Intake Systems you can find on the web:

www.knfilters.com

www.airaid.com

www.afefilters.com

Optimizing Fuel Quality and Delivery

The fuel system and its ability to properly deliver gas or diesel are critical to your vehicle's normal operation. Enabling your system to mix fuel with air and a spark to create perfect combustion every time in every cylinder of your engine requires a great deal of synchronization and cooperation with other systems.

The two primary types of delivery systems are fuel injection and carburetion, although all cars sold in the U.S. since 1990 have, by law, included fuel injection systems.

There is a reason fuel injection has been chosen over carburetion as the system of choice. Fuel injection systems manage the fuel delivery much more precisely and by doing so maintain higher operating efficiency throughout the engine's operating range.

Some older cars with carburetors are an absolute "bear" to tune to get a precise air to fuel mixture at the same level of performance as a fuel injection system.

Fuel Quality

By providing clean, high quality fuel to the engine, you are giving it a better chance to deliver its best efficiency. On the other hand, by delivering poor quality fuel, you are introducing lots of ways for things to go wrong.

Refined fuel, like gasoline, immediately begins to degrade the moment it is refined. Oil companies try to extend the potential shelf-life of gas with certain additives to maintain its ability to create combustion. Low octane fuels degrade faster than high octane fuels.

Think of gasoline, and its additives, like Nestlé's Chocolate syrup when added to milk. Once you stir it up in the glass, and you get a delicious chocolaty drink. But let it sit, and soon you'll find all the chocolate at the bottom of the glass. Like Nestlé's Chocolate and milk, you get much the same effect with gasoline and its additives.

In today's cars, your fuel injection system manages these inconsistencies much better to provide a more consistent air-to-fuel mixture than carburetors ever could. Even so, gas more than a few months old could start leaving gum or varnish deposits on the internal surfaces of the fuel system, possibly gumming things up.

Diesel is particularly susceptible to gumming up when it gets old or when it gets cold outside. These issues can be resolved by using a quality gasoline or diesel additive if the fuel in your vehicle is going to sit for more than a few months.

Stale diesel can also be susceptible to microbes that are attracted to the moisture in the diesel. Burning microbe infested diesel could clog the fuel pump or even the injectors. An anti-microbial additive is available if you intend to store diesel in a tank for a long period (for example, storing your RV with a diesel engine over the winter).

Another problem with low quality fuel is water contamination. Water in the fuel settles to the bottom of the tank where the fuel line pulls fuel from the tank. Water in the fuel can cause significant ignition issues and could even stall your vehicle. Liquid additives are available that mix with the water, breaking up the molecules sufficiently that the water passes through the engine with the fuel harmlessly and escapes in the exhaust as vapor. Water in the fuel can come from a gas station with a leaking underground storage tank or from water condensation in your gas tank.

How Fuel is Delivered to the Engine

Conceptually, fuel delivery is simple. Fuel is stored in a fuel tank, typically under the rear of the vehicle. Fuel is pumped from the tank via an electric fuel pump, through a fuel filter that removes particles and impurities, and then to the fuel injection system (or carburetor on older cars).

Did you know that *the gas pedal is really an air pedal*? When you press on the "gas", it opens up the throttle valve, which lets more air into the engine. Then the ECM or ECU (Electronic Control Module/Unit depending on who you talk to), your engine's brain, increases the rate of fuel being supplied to the engine.

The ECM carefully monitors the mass of air entering the engine, the amount of oxygen in the exhaust, and other parameters to meter out the precise amount of fuel.

The result is a very precise air-to-fuel ratio of 14.7:1 that is designed to burn most efficiently in the combustion chambers of the cylinders.

The fuel injectors themselves are simply electronic valves squirting a super fine mist of fuel into each cylinder. The finer the mist, the more complete the burn, and the more energy extracted from the fuel, given that all the other parameters of the combustion process are favorable.

As the ECM meters fuel to the fuel injectors, an electronic signal is sent to each injector, telling it when and how long to open to squirt the fuel mist into the cylinder.

The timing coincides with the opening of the intake valve and the downward movement of the piston on its intake stroke, which draws the fuel-air mixture down into the cylinder.

Fuel Evaporation and Pollution

Every year, hundreds of millions of pounds of pollutants are pumped into the air from the evaporation of fuel alone. This is not counting the emissions from the tailpipe. We're discussing fuel evaporating from the gas tank from other components of the fuel system.

The primary culprit is a leaky gas cap. Most vehicles over three years old, because of the aging of the seal, are in need of a new gas cap. The average vehicle with a leaking gas cap loses 30 gallons of gas per year. This same vehicle dumps 200 pounds of gasoline vapors into the air which combine with heat and sunshine to form Ozone.

The Gas Mileage Bible

It is estimated one out of five vehicles has a faulty or missing gas cap which accounts for the evaporation of 150 million gallons of gas per year in the United States.

Ozone high in the atmosphere protects us from harmful ultraviolet light. Ozone at ground level (we often call it smog) causes a great deal of health problems, deterioration, and unpleasantness. According to the American Lung Association, air pollution:

- Contributes to asthma, emphysema, lung cancer and bronchitis.
- Soils clothes, houses, and automobiles.
- Makes paint peel and discolor.
- Rusts and tarnishes metals.
- Reduces visibility and sunlight.
- Destroys plants and crops.
- Deteriorates statues, monuments and buildings.

Tips and Techniques for Optimizing Fuel Quality and Delivery

Solution #22   Clean or replace your fuel filter regularly

Having your fuel system deliver quality filtered fuel is critical to the proper function and optimization of your vehicle. Keeping the fuel filter changed within the schedule listed in your owner's manual will reduce wear and tear on other components in your fuel system. Varying fuel pressure or contaminated fuel can lead to various problems, including fuel injector failure or premature fuel pump failure.

Solution #23 **Buy the lowest octane grade of fuel recommended for your vehicle**

The best choice of putting fuel into your vehicle is to use the lowest recommended OCTANE (R+M/2) your manufacturer specifies for your vehicle. NEVER GO BELOW THE OCTANE level recommended for your vehicle! This can lead to engine damage or worse.

Many people feel that filling up with the highest grade (93 Octane) is the best way to get top performance from your vehicle. There have been many studies concluding higher octane gas nets little or no efficiency or performance gains, especially considering today's higher cost of the fuel.

Here's another idea that may make sense for you. If you get your gas from a busy gas station the gasoline will be much fresher (as it does get stale after a while) and it won't have time to collect water or other contaminants if the underground tanks are leaky.

Solution #24   Check your gas cap

A faulty seal on your gas cap can account for the evaporation of ***30 gallons*** of gas every year. In warm climates, or if your gas cap is completely missing, you could be losing even more! At today's gas prices, that's close to $100 a year!

A faulty gas cap adds a tremendous amount of pollutants into the air. Volatile Organic Compounds (VOC's) and Nitrogen Oxides, which escape from an unsealed gas tank, combine with sunshine and heat to form summertime ozone pollution. Each leaking gas cap is responsible for ***200 pounds*** or more of pollutants each year.

If your car is more than three years old, you may need to replace your gas cap, as the rubber seal will be worn and less effective.

It's easy to get a replacement gas cap. Most automotive stores carry them. Make sure to get a good quality cap by testing it to make sure it has a tight seal when you put it on.

Many states are now inspecting the gas cap seal during annual automotive inspections by pressurizing your gas tank and watching for excessive leakage around the cap.

Solution #25   Don't let fuel get stale

Contrary to popular belief, stale fuel will most likely combust reasonably well, unless it is so old (many months or even years) that it has started to gel. The biggest problem is that stale fuel will leave varnish or lacquer-like deposits and gum up the inside of your fuel system. There are additives for cleaning out those deposits that will reportedly even restore old, stale gasoline or diesel to a near-new state.

So, if you have fuel that has been in your tank for more than two months, it is possible it will burn inefficiently, leading to poor fuel economy and performance. It will also likely start leaving deposits by then. As a rule of thumb, if you have fuel more than two months old, be safe and use one of these additives or another quality brand:

- Sta-Bil, manufactured by Gold Eagle Company (reportedly also restores old, stale fuel to a useable state)

- Nalco Winter Thaw Emergency Diesel Fuel Treatment (for diesel that has already gelled – emergency use only)

Beware of Fuel Pills, Liquids or Other Additives!!

If you scan the internet for fuel saving products, you'll get hammered with ads and web pages promoting many different fuel additives that are supposed to improve your miles per gallon. The gimmicks and shams are only going to get bigger as the price of fuel continues to rise. Here's our take on them:

1. We generally only use fuel additives when we get bad gas, or as a preventative measure to ensure we don't have water or other impurities in our gas.

2. We've had limited success getting better incremental mileage as a result of using these additives.

3. It is very difficult to find real test results by a certified testing authority that says fuel additives work <u>consistently</u>. (Our experience shows that you get a temporary increase in mileage because of the cleaning effect of the additive, but no other long term gain).

4. The Environmental Protection Agency routinely tests fuel saving products to see if they really work. As of this writing, they had not tested fuel pills yet, but most other gadgets and gimmicks have been found not to work.

5. The Federal Trade Commission routinely shuts down many of the companies that sell these products and make false claims.

NOTE We are constantly updating our base of knowledge. If we feel comfortable with a new product, we will let our readers know. Make sure to sign up for our mailing list on **www.gasmileagebible.com** or on **www.vitalshift.com** so that you can be notified of new information and any new recommendations we make.

Solution #26 Clean and service fuel injectors as recommended by your service manual

As we had discussed in the solutions before this, having very clean fuel injectors, as well as fuel, is a big deal to getting the performance you are looking for from your vehicle. Without your injectors working at peak efficiency, and precisely delivering fuel, the combustion process inside your engine will deliver less than 100% of the energy it was designed to provide.

If you have injectors that are fouling or not delivering fuel exactly as needed in the cylinders, the ECM WILL retard spark timing, falling out of the optimized program table.

Your vehicle's computer is programmed to operate within a certain range of conditions. When conditions fall out of the programmed range, the computer attempts to adjust whatever it can to get back into its operating range. That's why it adjusts the spark timing as described above.

The ECM, also being fed sensor readings from the mass flow sensor and oxygen sensors, may cut fuel completely to that cylinder if it cannot keep up with the injector variance.

Your ECM maintains a delicate balance between drivability and outright performance. Keeping the injectors clean and dirt free helps maintain that balance.

We keep saying this is a big deal and it is. If your engine is out of tune and operating poorly, it could be costing you as much as 40% more to drive your vehicle. So, when gas is $3.00 a gallon, you are effectively paying as much as $4.20 a gallon when your engine is not tuned and operating properly.

Yes, yes, we know. It's hard to remember to do any maintenance on your vehicle until something actually breaks and gets your attention. THAT's when MOST PEOPLE remember to maintain it. But, if YOU do things differently, and choose to proactively maintain your car or truck, you will be well on your way to saving many hundreds of dollars a year.

Don't let your fuel tank get too low!

As your fuel level drops to the bottom of the tank, left over sediment and dirt that have settled there get sucked into the fuel line. At best, your fuel filter will capture all that gunk, but will die an early death as it gets clogged up.

A clogged up fuel filter restricts the flow of gas and will adversely affect performance, up to the point of fuel starvation and engine shutdown. Worst case, some of the sediment gets through the filter and clogs one or more injectors. This is a bad thing!

As a rule of thumb, don't let your tank get below about one quarter full and you'll never need to worry about this.

Optimizing Fuel/Air Combustion, Ignition and Timing

How it All Works Together

Sensors all over the engine watch over and monitor the parameters affecting the air fuel ratio (AFR). The outputs from these sensors feed the ECM which controls the precise ratio of air and fuel entering the combustion chamber.

We mentioned how the driver actually controls the air flow into the engine by pressing on the gas pedal. The ECM monitors air flow, the amount of oxygen in the exhaust system, the mass of air in the air intake system, and the several other parameters.

The ECM is programmed to establish the "perfect" ratio of 14.7:1. That is, it adjusts the fuel delivery so that for every 14.7 units of air, one unit of fuel is delivered. This number has been determined scientifically and through years of experience with internal combustion engines to be the air fuel ratio at which gasoline is burned most efficiently.

If there is more air in the mixture, then it is said to be running lean. If there is more fuel in the mixture, then it is said to be running rich. Multiple sensors throughout the air intake, fuel, engine, and exhaust systems monitor flow, temperature, pressure, and other parameters to feed to the ECM for maintaining a just right mixture; not too rich, and not too lean.

Compression of the perfectly mixed air and fuel being injected into the top of the cylinder is the next step in the combustion process. The mixture was sucked into the cylinder by the down stroke of the piston. Now the piston is moving back up and compressing the air fuel mixture into a very compact area, and exerting very high pressure.

As the piston nears the top of its stroke, an electric charge is sent through the spark plug and a spark is introduced to the highly volatile mixture. The mixture ignites as the piston is reaching the top of its stroke.

The explosion is timed to reach its maximum force just as the piston reaches the very top of its stroke and begins the return journey to the bottom of the cylinder. The explosive force of the combusting air fuel mixture drives the piston down, creating the force powering your engine and moving your vehicle forward.

Controlling the dance of the ignition system is the distributor. The distributor is the mechanism keeping things synchronized and telling which spark plug is to receive the next spark.

Since the cylinders fire in a specific order, the distributor is designed to send the electric signal to the next cylinder in line for ignition. The order in which the cylinders fire is calculated to evenly distribute the force of combustion in each cylinder and results in a smoothly running engine.

The ignition coil is a source of very high voltage (as much as 100,000 volts!) which gets sent to each spark plug in turn. The distributor routes the signal to the proper spark plug.

The distributor timing can be fine tuned to send the spark sooner or later for any given spark plug. This allows the ignition spark to be timed so that the air fuel mixture is ignited at just the right moment, and the maximum force of the resulting explosion occurs just as the piston reaches top dead center (TDC).

Then the entire force of the explosion is expended on driving the piston down. This is appropriately named the *power stroke*.

If ignition were to occur too soon, the explosion would peak before the piston reaches top dead center, and there would be a detrimental force actually counteracting the forward momentum of the piston.

You can bet this has a negative effect on engine performance and fuel economy! This is one reason a poorly tuned vehicle can get as much as 40% lower gas mileage than it should be getting.

If the timing of the spark is too late, the explosion peaks after the piston has already started on its downward stroke. The force of the explosion is less because it is no longer tightly compressed at the top of the stroke, and the piston is not driven with as much force as possible.

Again, a less than optimal performance and fuel economy hit on the engine. You can see proper timing is crucial for a well run engine. The best way to maintain proper timing is to ensure your car is properly tuned by a professional mechanic with the right diagnostic tools.

Choosing the Right Spark Plug

Your ignition system is the heartbeat of your engine. Having the right spark, at the right time, in the right sequence, is central to what the engine needs to operate efficiently. Spark plugs have been around, in one form or another, since the creation of the first single cylinder internal combustion engine.

We're pretty sure the complexity of modern engines would just flat floor the creators of those first engines. There are engines today producing greater than 500 HORSE-POWER yet can attain over 20 miles to the gallon of gas!

That is progress and that is evolution. Let's discuss some new technologies you can use for your older engine. Did you know you could borrow the latest spark plug technology from modern engines and use it in your older vehicle today?

Spark plug metallurgy and usage has varied from steel, then to aluminum, silver, gold, nickel, platinum and now iridium. The graph below shows the characteristics of the best all-around plug for your vehicle. How to interpret the graph below:

- The higher the melting point of the plug, the longer and cleaner it will burn throughout the temperature range being used in most factory engines today.

- The higher the strength of the plug the longer it will last under sustained load and usage.

- The lower the electrical resistance of the plug the more energy passed through it and into the combustion chamber (the lower the *loss* of energy).

- The harder the metal the more durable the plug will be for its lifetime.

Here is a table showing the melting points and other characteristics of commonly used metals in today's spark plugs:

	Iridium (Ir)	Platinum (Pt)	Nickel (Ni)	Gold (Au)	Silver (Ag)
Melting Point (°F)	4449.2	3216.2	2647.4	1945.4	1760
Strength (1000's PSI)	159	19.9	96.5	18.5	18.5
Electrical Resistance (µΩ♦in)	2-3/32	4-11/64	2-11/16	29/32	5/8
Hardness (HV,68 Degrees-F)	240	40	160	25	26

If you look at all of these traits to measure spark plugs the obvious choice is the iridium. It is now recommended by top manufacturers for maintenance intervals of 60, 80 or even 100 thousand miles! Remember, optimization of your engine can also mean optimizing the *time you spend* on recurring maintenance where possible.

So, our basic recommendation is to *get the highest range plug that will work well with your vehicle.*

Due to improvements in design, materials, and technology, some 2005/2006 vehicles have a 100,000 mile maintenance interval across the board for the entire engine!

That is a leap in technology! If you ask ANY car parts salesperson or anyone who has been working on cars since the 1970's they will tell you the jury is still out on trusting this very extended interval.

One item worth noting. Most new car manufacturers with a 100,000 mile service interval are using plugs with the higher-end metals in their design and construction. An older car with 80-100,000 miles on it may have a bit of an issue with performance with these higher end plugs.

If your older engine consumes a bunch of oil then you might be better served using the next lower heat range to get a plug that will not "foul-out" after 1,000 to 10,000 miles for oil and "blow-by" deposits getting into your cylinder.

This had happened to me (Kenny) 2 years ago on a vehicle I had where I thought I could put these higher end plugs into it without a problem. The car ran GREAT for 500 miles and then began to foul out. I took the plugs back to the parts store because hair-line carbon deposits had formed on the electrode space of the plug, fouling it out.

The Father of Spark Plugs!

The first known spark plug was invented on February 2, 1839 by Edmond Berger. France dominated the spark plug market in the early 1900s, but supplies were limited in the United States. Albert Champion, a renowned French bicycle and motorcycle racer, came to America to compete in a series of races.

Parts were for his engines were hard to find so he made his own. To cover his costs of racing, Champion made spark plugs and sold them to friends. Champion's love of motors slowly turned towards automobiles and he eventually founded **Champion Ignition Company** for the manufacturing of spark plugs.

After a dispute with his investors, he lost his company, but the investors continued to manufacture spark plugs under the Champion name. So, Albert Champion began a new company called the **AC Spark Plug Company**.

The AC Spark Plug became a division of General Motors and eventually evolved into the company we know today as **AC-Delco**. To this day Albert Champion's name lives on with every AC and Champion spark plug made.

Here are some more interesting AC Spark Plug Facts:

1927- AC spark plugs were used in Charles Lindbergh's plane in his trans-Atlantic flight.

1932- Amelia Earhart's plane was equipped with AC spark plugs when she made her trans-Atlantic solo flight

1969- AC igniter spark plugs were used to fire the second and third stage rocket engines that took Neil Armstrong, Buzz Aldrin and Mike Collins to the moon.

The plug was sent back to the manufacturer (they really wanted to know why their $8.95 spark plug had failed). The manufacturer said they analyzed the compound on the plug, and determined the cylinder had too much oil in it and had failed the plug. I dropped down a grade and everything was ok after that.

While I did not have the "raw" performance I was looking for from the best plugs available, I did have an incremental increase is performance and gas mileage.

To learn more about how the combustion, ignition and timing work in your vehicle, please visit *www.howstuffworks.com*.

Keep your vehicle properly tuned

Make sure you keep your vehicle properly tuned by a professional mechanic who has access to the right diagnostic tools. He will make sure your exhaust gas emissions are in balance and will ensure your ignition timing is set properly. An out of tune vehicle can easily result in a performance and fuel economy hit of 10 to 40%!

Tips and Techniques for Optimizing Combustion, Ignition and Timing

Solution #27 **Use high quality spark plugs and ignition cables**

To extend your service interval and fuel mileage then there are a few alternatives you may want to consider. Here are some spark plug replacement alternatives for your engine:

- www.ngk.com
- www.acdelco.com
- www.boschusa.com/AutoParts/SparkPlugs

These manufacturers' spark plugs are original equipment replacement for most engines today.

Spark Plug Case Study

Here is a case study to demonstrate changing of the plugs as a step in optimizing a gas guzzler for better fuel economy.

We wanted to replace the original plugs in a 1997 Lexus LX 450 (Basically a Toyota Landcruiser) with 82,000 miles on it. We went to the NGK web site since the vehicle came with NGK-R plugs, the high end model at the time of manufacture.

We found a nifty internet product configurator that gave us low, medium and premium choices but also gave us choices for other accessories! But, there was NO choice for a replacement to the NGK-R! That model is now available only to dealerships as factory replacements. So, we went with latest premium brand from NGK, the NGK-IRIDIUM plugs. They have a lower voltage resistance that allows a larger spark with a much smaller electrode.

Since the vehicle had been maintained flawlessly, the Iridium plugs were a good choice and caused no problems. Had there been excess oil in the cylinders or other signs of poor maintenance, we would have likely chosen a lower grade of plug. Since that change, we have seen a one to two MPG increase in gas mileage on the highway!

Solution #28 **Keep your engine tuned and well maintained**

This is an area many people and agencies, including the EPA, speak of repeatedly. There must be some validation of this as most people only loosely adhere to the manufacturers recommended maintenance!

Let's be very clear about this. Keeping up on your maintenance is not a suggestion; it is the first *requirement* to even embarking on *any* kind optimization. In scientific terms, we call this the *baseline*.

<u>A poorly tuned engine can cost you up to 40% of the gas mileage you should be getting!</u>

Before you do anything we suggest, we highly recommend all other manufacturer recommended maintenance be up to date.

For example, changing your air filter during the required intervals is a must. Performing all of your required fluid and filter changes is a critical and absolute must! Did you know that NOT changing your $19.95 fuel filter could result in a $500.00 to $1,000.00 repair of your fuel injectors and engine management system? Believe it.

We want you to achieve a greater awareness of what your car is doing and how it is responding to changes, conditions and even your driving habits.

This awareness could make the difference between having a great vehicle ownership experience and an ownership nightmare! Most people are not even aware their vehicle is out of balance and not operating at peek efficiency!

We recommend you take 30 minutes to re-acquaint yourself with your owner's manual. There is a wealth of information in there regarding your vehicle. Automakers spend millions of dollars putting together the information they think you need to maintain your vehicle.

This is the first place to start. After reviewing your owner's manual you need to track down where you are on your vehicles overall maintenance. Do you know where you kept that last receipt for the oil change you did in the garage in say 2002 but can't remember? Do you keep any records? Do you log what and when you perform any maintenance on your cars? Would you rather buy the pine-scented tree hanging from your rear-view mirror rather than change your oil?

Only you have these answers. Maintaining your engine to the manufacturer's recommended specifications is only the *start* to realizing the benefits of this book. We want you to have complete situational awareness of your engine, transmission, tires, brakes, electrical, and air-conditioning maintenance before trying anything in this book beyond what the manufacturer recommends.

If you don't have that awareness, then pull out your owner's manual, get what receipts you have, write down what you think you have done, and get the advice of your local trusted mechanic or your dealership if necessary.

Establish a maintenance baseline. Form the habit of maintaining your vehicle regularly and log everything you do. This will save you many dollars in the future and give you peace of mind now.

Optimizing Engine Exhaust

Back pressure is the arch nemesis of your engine exhaust system. Just as your engine likes to breathe in lots of fresh cool air, it likes to be able to exhale the used up air and gases from its combustion process.

If there is too much back pressure, the engine must work harder than necessary to expel its exhaust. Additionally, a condition known as "engine-heat-soak" or "heat-crawl" will occur where trapped heat will travel back up the exhaust, through the catalytic converters, up the manifold, and into your entire engine. Heat is the robber of performance and mileage in any engine. Reducing back pressure reduces heat in most cases. This help preserve engine performance.

Because of emissions laws and other carmaker restrictions, most cars come equipped with overly restrictive exhaust systems .

One of the quickest and easiest ways to improve engine performance and fuel economy is to provide the engine with all the air it needs and let it expel its exhaust gases effortlessly. Improving your air intake and exhaust can result in some amazing gains. Let's first identify where the exhaust gases come from.

The exhaust process begins in the cylinder. Once the air fuel mixture has exploded and driven the piston down on its power stroke, the piston starts back up again.

This time, its role is to push out the remaining exhaust gases through the exhaust valves and into the exhaust manifold. From the manifold, gases travel through the exhaust pipe to the catalytic converter, through the muffler, and out the tail pipe.

The purpose of the muffler is obvious; it quiets the sound of the exhaust. We're sure you've heard the sound of a car that has lost its muffler before – obnoxiously loud! You may not know the muffler actually adds to the restrictive force that is "felt" by the engine.

The role of the catalytic converter is somewhat mysterious and not many people understand its purpose or importance. We'll discuss the amazing yet simple technology found in the catalytic converter in a bit.

Restriction in the Exhaust System

Exhaust gases travel quite a distance from the exhaust manifold to the tail pipe. Every inch of the way there is a resistive force.

Imagine you have an overflowing bathtub in the house. You can't turn off the water, the drain is clogged and you are afraid of damaging the floors. The only way to keep the water from damaging the floor is by draining the water through a garden hose you brought into the house.

Quickly you find out the tub is filling faster than the hose will allow it to drain. The only way to speed it up is to get a bigger hose or to force the water through the hose under pressure. Now suppose you could seal up the tub somehow (including the faucet, which is still running!).

The faucet represents the air intake of your engine, the tub represents the engine and the garden hose is your exhaust. Don't get too excited – there are no plans yet to build a car that lets you bathe on the way to work....

When the tub fills up, your seal causes it to build up pressure. As long as the seal holds, the pressure will force the water to drain faster through the garden hose – at the same rate it is entering the tub through the faucet.

That is how much your engine has to work to "push" the exhaust gases out of the restrictive exhaust systems on many of today's cars.

Now, what if you could replace the garden hose with a fire hose? Would you need to force the water out under pressure? Probably not. The hose would be big enough to allow the water to drain of its own accord. Your "engine" (the bath) need not work so hard under pressure to force the water out.

It makes sense then, that by adding an exhaust system with larger diameter pipes and lower restriction through the muffler and catalytic converter, you could decrease the back-pressure "felt" by the engine. Hmmm.

Do you think there is a big market for those kinds of modifications to your vehicle? You betcha! We'll discuss that more in a bit.

Pollution and the Role of the Catalytic Converter

One of the reasons automakers build such restrictive exhaust systems on our vehicles is because of EPA pollution laws. They have been under the gun for many years to improve the efficiency and emissions of the vehicles they sell.

The internal combustion engine is far more efficient than it was in years past. Vehicles emit fewer pollutants than ever before, but still put out a significant amount of damaging emissions, especially when multiplied by the millions and millions of vehicles in service today.

Some will tell you the internal combustion engine has reached the optimum amount of efficiency that can be squeezed out of it.

Environmentalists discuss the effect of greenhouse gases and global warming, and the role automotive emissions play in creating those problems.

So what gases or pollutants in the emissions of a modern vehicle are so bad? Let's take a quick look…

The emissions from your vehicle include primarily water vapor (H_2O), nitrogen (N_2), and carbon dioxide (CO_2). These are pretty benign gases and cause no harm directly to humans.

Carbon dioxide, however, is a so-called greenhouse gas that contributes to global warming.

The more damaging (and regulated) gases in your emissions include carbon monoxide, hydrocarbons (or volatile organic compounds (VOC's)), and nitrogen oxides (NO and NO_2, together called NO_X).

Hydrocarbons are essentially unburned, evaporated fuel. Carbon monoxide is a dangerous gas that is fatal to humans, even in very small doses. Nitrous oxides are a major component of smog and acid rain and irritate human mucous membranes. They also react with carbon monoxide and hydrocarbons to create ground level ozone, another major contributor to smog.

These last three emissions are the most dangerous and the most regulated. The catalytic converter was designed to help reduce the levels of these three gases in our exhaust systems.

Someone came up with a simple yet ingenious device called the catalytic converter. Modern three-way catalytic converters include two types of catalysts embedded in a honeycomb structure or ceramic beads that act on the three dangerous emissions gases.

First, the gases pass through the reduction catalyst which converts much of the nitrous oxides to nitrogen and oxygen. Then the exhaust gases pass through the second stage, the oxidation catalyst, where the unburned hydrocarbons and the carbon monoxide are essentially burned or oxidized.

This is possible in part because the catalytic converter heats up due to its proximity to the engine and the temperature of the exhaust gases. The other enabling factor is the existence of oxygen in the exhaust. One or more oxygen sensors between the engine and the catalytic converter monitor oxygen level.

The oxygen sensor sends its report back to the ECM which can adjust the air fuel ratio to allow sufficient oxygen in the exhaust stream to support the oxidation process.

Catalytic Converter Woes

While the catalytic converter does a very good job of reducing the dangerous emissions, it is not perfect. The biggest problem is that it needs to be hot to be most effective. When you first start the vehicle in the morning, the catalytic converter is practically useless for a minute or two (or longer when it's cold out) until it warms up.

Solutions that have been suggested include moving the converter closer to the engine so that it heats up more quickly. Automakers don't like this solution because at the higher operating temperatures, the catalytic converter would burn out more quickly – and they are very expensive!

Another solution is to use an electric heater to warm up the converter before you start the vehicle in the morning. Most people would not want to wait several minutes in the cold weather for the converter to warm up before starting their engines, and the electrical systems on our vehicles don't have the power to heat them any faster.

We can do our part to minimize these bad emissions by ensuring our vehicle is in top notch condition with a good tune-up and a number of the optimizations listed in this book. An optimized engine burns fuel more efficiently and leaves a minimum amount of exhaust emissions.

Measuring Exhaust Gases to Optimize your Engine

Guesswork was the name of the game in the old days when it came time to tune your engine. There were not very precise methods to adjust your timing and other aspects of your tune-up. Adjusting your carburetor was always a big guess.

Now with the advent of computerized fuel metering and knowledge of the "perfect" air fuel ratio of 14.7:1, we are sure the computer is feeding the right amount of fuel and our engine is running efficiently, right? Well, maybe. There are a number of things that could go wrong, from a malfunctioning sensor or injector, to computer chip data corruption.

On all new automobiles, there is at least one sensor in the exhaust monitoring the amount of oxygen (O2) in the exhaust gases. This sensing of the gases coming out of the tailpipe dictates the pattern and the mapping table the ECM uses to control engine operation.

The ECM controls the mass flow air sensor, ignition, fuel delivery and timing to maintain the perfect air fuel combination so your engine runs as efficiently as possible throughout the entire operating and temperature range.

If your exhaust gases are NOT what the ECM is looking for then it immediately degrades to other, less efficient firing patterns., It keeps trying until it finds a pattern that sufficiently maintains engine operation without causing any engine damage.

If it senses parameters outside its programmed operating range then it will flash the check engine light, or revert to what some manufacturers have called the "limp home mode".

In this mode, the computer picks the minimal amount of cylinders it can use to just get the vehicle home. If you ever see this condition, just know you will be having a date with your dealership!

So, it is very important to keep your oxygen sensors in the exhaust in the best possible working condition! Most people aren't even aware these critical sensors exist!

This is a critical point! If you can't monitor it, you can't manage it! If you have a defective sensor you could be driving a vehicle well below its operational potential, with correspondingly poor fuel economy.

The moral to this story is to make sure that your oxygen sensor is always in perfect working order! Your mileage and pocket book depend on it!

A sure fire way of optimizing your engine combustion and retrieving the most energy from your engine is to measure the emission gases your vehicle is putting out. There are sensitive exhaust gas analysis tools to tell you exactly how efficient your engine is running. Every good mechanic's shop should have one. When combined with other diagnostic equipment, today's technicians have an unprecedented amount of visibility into the inner workings of your vehicle's engine.

Tips and Techniques for Optimizing Engine Exhaust

Solution #29   **Tune your car up and get an exhaust gas analysis**

Pollution compliance in most major cities is driving a trend that all cars within certain age brackets be tested for emissions. This inspection can give you a *wealth* of information on the health of your car, *right now*.

Most inspection stations charge $12.00 to $20.00 to perform an exhaust gas inspection of your car. This can give you a very clear picture of how your vehicle is performing at a very reasonable price.

Now, understand this other element. There are reporting services such as CARFAX and EXPERIAN taking the data from state licensed inspections, building a profile of your vehicle, and keeping a historical record.

Your failure to pass the inspection will be recorded on your vehicle profile with these services forever! Be aware of this before going into this inspection stations!

You can also pay (probably a bit more) for a similar analysis to be conducted by a reputable tune-up shop and not have your less than optimum information picked up by the reporting services.

Solution #30 Make sure your Oxygen Sensor is Working Properly

Fixing a serious maintenance problem, such as a faulty oxygen sensor, can improve your mileage by *as much as 40 percent*, according to the Department of Energy.

Remember, it is the Oxygen Sensor sending signals back to the ECM (the brain in your engine) that helps determine the air fuel ratio that your engine is receiving. If it is sending the wrong information back, you could literally be pumping excess fuel out your exhaust!

Detecting a faulty Oxygen Sensor is something your local dealership or mechanic should do with the engine diagnostic equipment they use. Once a faulty oxygen sensor is identified, changing it out is something most home mechanics can do themselves.

Solution #31 Install Lower Resistance Exhaust Components

We've discussed letting the engine breathe more easily. Meaning, we want to minimize the air intake restrictions as well as the exhaust outlet restrictions. We've already discussed methods for reducing air inlet restrictions. Now we will discuss doing the same for the exhaust, something we call reducing backpressure.

Keep in mind this is a much more expensive and long term solution. If you intend to keep you vehicle for at least two or three more years or you drive full time, this may be a solution for you. If you do not intend to keep your vehicle for long, don't waste your money on these upgrades.

If you do take the plunge and modify your exhaust system, you will notice significant improvement in both fuel economy and power. This is a highly recommended solution for those who haul long distances. For instance, if you pull a fifth wheel or travel trailer, or drive a work truck extensively, it may be a good idea to open up your exhaust system to let your engine breathe better.

There are several options for reducing backpressure in your exhaust system. You can replace one or more components, or you can replace the entire system. Here are a range of options:

- Replace your muffler with a high-flow model
- Replace your catalytic converter with a high-flow model
- Replace your exhaust manifold with headers
- Replace your exhaust pipe with a larger diameter pipe

OR

- Replace the entire system with a tuned, high-flow exhaust system
- Replace the entire system with a dual exhaust high-flow exhaust system

Augmenting or replacing exhaust systems on today's vehicles is relatively expensive and in some cases may not provide the return on investment if you just count gas mileage savings as your payback. If you are looking for more power, a different sound or just bragging rights, you may be able to justify the expense.

However, if you have an older or high-mileage vehicle (90-100,000 miles) then you may want to strongly consider replacing your catalytic converter with a higher-flowing, yet still effective, unit.

This change can result in improved fuel economy if you have followed our advice, achieved a maintenance and fuel mileage baseline and still not gotten the fuel economy you were looking for.

In today's newest cars, manufacturers are spending a great deal of time and resources to create an exhaust note, tone, burble, pop or anything else it wishes to tune into the sound of the exhaust. They want to create an ambiance or "style" for different vehicles to appeal to certain markets and drivers. This tuning can marginally affect performance in exchange for a quieter or unique exhaust sound.

Augmenting your system with a performance tuned exhaust can result in a remarkable increase in power and efficiency, but at the cost of more noise. Mufflers work by converting the energy in sound waves into heat. With a higher performance exhaust system, less sound energy is converted to heat and passes directly out of the tailpipe.

Many companies out there tout their performance exhaust systems. Just a few we have used with positive results are:

www.borla.com

www.gibsonexhaust.com

www.edelbrock.com

www.flowmaster.com

Each of these is a leader in its area and recognized for having specific tuning options for your specific need. The moral of the story here is that they spend millions of dollars each year in the pursuit of getting more performance out of your engine.

Be warned, some of these exhaust options will net you a performance gain. What you may give up in exchange is a bit of the luxurious and quiet sound your stock exhaust used to make.

Here's an extreme example of a noisy performance exhaust system. In a 1992 LT-1 Corvette, someone we know installed a top-end exhaust system from Flowmaster. The exhaust was so loud the owner had to install more sound-deadening material in the vehicle in order to drive the corvette without being distracted by the noise. Sometimes you can go too far. Be aware and make the right decision for you.

Optimizing Lubrication and Cooling

Optimizing lubrication and cooling in your engine are essential to an efficient power plant. We've covered these topics in Chapter 3 and provided some recommended solutions.

Probably the most important thing you can do to enable efficient engine operation is to use synthetic oil in your engine. The synthetic oil will allow your engine to operate much cooler, will provide superior lubrication and will last much longer than petroleum based oil.

Optimizing Mechanical Health and Tuning

The vast majority of vehicles on the road today over one or two years old are probably significantly out of tune and under-maintained. They are probably operating far more inefficiently than they are capable of. What most people don't realize is the cost of this inefficiency in terms of gas mileage.

The problem is when a car is more than a year or two old, it has likely fallen out of tune, taken a few lumps and bumps, and has become less efficient. The engine parts get broken in and then begin to wear, spark plugs wear out, fuel injectors get gummed up, and so on.

The same is true with mechanical components such as the suspension, wheel assemblies and a dozen other systems. This is normal for a vehicle – to show the signs of use. Some get driven harder than others (how many curbs and pot-holes have been personally introduced to your wheels?). Remember even a small misalignment of your vehicle's front end and tires can result in a dramatic increase in rolling resistance.

But all vehicles need regular attention if we want them to be efficient and get better mileage. The problem is typically US!

We tend not to have time to take care of our vehicles as much as we should. We always have something more important demanding our attention. With the exception of Kenny (☺), every other driver on this planet could probably pay a little more attention to the regular maintenance and upkeep of our vehicles.

Mark your important maintenance dates on your calendar, or in your electronic organizer. Keep a log in your glove box that tells what you did and when.

> **Set a Date with Your Car**
>
> If you use an electronic organizer such as Microsoft Outlook or the software on your Palm Pilot, put the critical maintenance dates for your vehicle in your calendar. Set it so that you get a reminder when an important maintenance event is due. Schedule it just like you might schedule a business meeting or a lunch date.

One of the most important results of a good tune-up is a more efficient combustion process. And, the combustion process must be efficient to get good fuel economy. That is when fuel is burned most completely and the maximum amount of energy is extracted from it.

Then, there will have been a relatively small amount of energy lost to friction, mechanical load and heat in a well tuned and running engine so the maximum amount of power gets to the wheels to move the vehicle forward.

Perhaps even more importantly, this engine will likely last significantly longer and require less maintenance than an under-maintained and out of tune engine. Please review the recommended tips and techniques in Chapters 3 and 4 for suggestions on how to keep your engine tuned to its maximum efficiency.

We've discussed ways to increase the efficiency of your engine. Before that, we discussed ways to minimize the loss of energy in your vehicle.

Next, we will discuss the incredibly important role your driving habits have on the gas mileage equation.

Tips and Techniques for Optimizing Mechanical Health and Tuning

Solution #32   Follow a regular maintenance schedule

You know that little book in the glove box? The one that comes with your new car or truck? The one you use to write down your family's fast food order on the back cover? Yeah, that's the one.

Well, these books all have a recommended maintenance schedule somewhere in there that includes oil changes, tune-ups, tire rotations and service check-ups with the dealer. They are a wealth of information, and many even give you a place to write down what you did at each recommended service interval, so you can track it.

We have only one thing to say about these cute little books.

USE THEM!

Maintenance Tracking

If you are really into tracking your mileage, trying different tips and techniques in this book, logging your results, and looking for things that really work for you, get your own log book.

You can get automotive log books at most variety stores, or you can use a simple half-size student spiral notebook that you can keep in your glove box with a pen.

We also discussed programs you can use on your hand-held organizer back in Chapter 2.

Kenny Joines & Ron Hollenbeck

Chapter 5: Alternative Fuels and Our Environment

Beyond gasoline or diesel – alternative fuels

The Department of Energy has listed a couple of alternatives to gasoline for vehicles in the United States. Did you know your newer (year 2000 or above) GM, Ford or Chrysler vehicle may be a flex-fuel vehicle (FFV)? That's right! For many gasoline fueled cars, there is an alternative fuel - a blend of gasoline and ethanol - you could use in your vehicle today. That fuel is E85 ethanol. It is 15% regular gasoline blended with 85% grain alcohol made primarily from corn in the US.

GM is claiming to already have 1.5 million E85 capable vehicles on the road today! It is not well known they have been selling these FFV's for several years now. Now GM is conducting an all out marketing blitz touting its current penetration in the "mainstream". The funny thing is, there are estimated to be fewer than 200 pumps dispensing E85 today in the entire USA. Even though there are so many E85 capable vehicles on the road, we must get the supply-side issues worked out before any wide-spread adoption of E85 can be realized.

There are few gas stations selling E85, so the biggest problem is finding somewhere to buy it. Here is a website where you can find the closest alternative fueling station:

www.e85fuel.com

Depending on where you live in the country, you may have lots of stations near you, or you may have none. Blended gasoline has been around for many years yet we have not yet adopted it very well.

There is some sacrifice in performance from using blended fuel, because of lower energy content. But that is yet another reason for us to create this book for you so you can maximize the efficiency of the fuel you use!

E85 flex-fuel contains approximately 28% of the energy content of regular gasoline. Many people are inferring there will be a directly proportional decrease in gas mileage when using E85. The expected mileage decrease for E85 flex-fuel is 25% to 30% below what "normal" gasoline gives you. According to the paper, "E85 and Energy Content" located on the National Ethanol Vehicle Coalition (NEVC) web site, www.e85fuel.com, however, you actually get better mileage using E85 than you might expect. They are claiming you get only 5% to 12% lower mileage than you do with regular gasoline.

When we put on our thinking caps and pondered this for a while, we realized there might be something to it. Remember, the internal combustion engine only uses 20% of the available energy in a gallon of regular gasoline. If the combustion of E85 is somehow more efficient, it might be possible to go further on less energy content. I guess we'll know more as people begin using E85 in quantity and we get some real world results to report.

E85 is currently less expensive than regular gasoline, so the bottom line to all this is that worst case you will spend the same amount per mile whether you use gasoline or E85. Best case, if E85 remains a better buy, and you really get better gas mileage than people think, you will be saving money for every mile you drive with E85.

The great thing about the flex-fuel vehicles being sold from GM and Ford since 2000 is that you must do nothing to make the switch to E85! Your FFV SUV or truck automatically senses this change and recalibrates the ECM to properly manage the fuel being introduced into it! Even better, you can use any combination of gasoline and E85 without having to worry about emptying your tank of gasoline before filling with E85, or vice versa. You can fill up with whatever is available any time you need to.

Using the tips and techniques provided to you in this book makes the switch to these blended fuels even more worth your while since you will be helping the environment and your pocketbook while you save gas.

Diesel and Bio-Diesel

In the past few years there have been discussions regarding diesel powered vehicles and diesel being an alternative to gasoline. Only in the past 3-5 years have US vehicle manufacturers given diesel a real look with development and engineering dollars (since an earlier attempt to market and sell diesel powered vehicles in the 1970's – a big flop).

In Europe 35 to 40% of all passenger vehicles sold are diesel, compared to a small percentage in the U.S, including mostly diesel pickups and a few SUV's. Volkswagen and a few other European auto makers are beginning to sell small diesel cars in the U.S. again, too.

In France and Austria, greater than 60% all passenger vehicles sold are diesel! The popularity of diesel in Europe is primarily based on the availability of cleaner diesel fuel, with a much lower percentage of sulfur. Americans think of the loud, smelly diesels of the 1970's and 80's sold here, and have a low opinion of diesels in general.

Diesels, by nature, are 25% to 30% more efficient than gasoline internal combustion engines. Not only do they get far superior gas mileage, they have an exceptional amount of torque (which equates to more pulling power and get-up-and-go) and they last longer. Today's diesels are stronger, quieter, and more fuel efficient than ever before.

The U.S. has mandated, beginning in 2006, diesel fuel formulation must be much cleaner and lower in sulfur and other pollutants – comparable with the diesel in Europe. Perhaps, with cleaner diesel fuel, Americans will be more open to buying diesel powered vehicles in the next few years.

There is an alternative diesel fuel known as bio-diesel. Did you know any diesel vehicle could burn a very clean alternative to "normal" diesel, and is for sale around the country? It is called bio-diesel because the "base" from which it is refined from comes from corn and soy oils, and not petroleum.

Did you know that the United States produces as much, or more, of these two crops than any country in the world! Did you know that for many, many years the United States government has subsidized our American farmers to grow less of these crops?

How can this be? If we have the supply, the demand, the means, the motive and the opportunity to create a situation to be much *less* dependent on Middle Eastern sources for our oil, don't you think we should take advantage of it?

The answer is the same as with most things in our country; politics and money. With current bio-diesel sourcing and production capacities, bio-diesel is selling at $3.00 to $3.10 per gallon. When gas prices reach a similar level (as they are now), the mass production and distribution of bio-diesel begins to make more economic sense. Everyday drivers of diesels (Like Ron with his Ford F-250) are more willing to fork over the premium price of bio-diesel, assuming it is available in their areas.

Other Alternatives

There are also alternative fuels being used and considered still under development. If you look closely at your city's vehicles and even your bus transit systems you will notice there may be some trucks, cars and buses being powered by E85, natural gas, or propane. That is right! We are dipping our toe into the alternative fuel pond as a nation!

There are some challenges with alternative fuels. Diesel powered vehicles have extra filters and a bit of a different maintenance schedule. Natural gas and propane powered vehicles do not have nation-wide access to fuel as we have with gasoline and diesel at the local corner gas station.

There are still many infrastructure and logistics issues. For example, the packaging to house and deliver the fuel in these forms is still in flux and yielding to the initial design for gas-powered cars because of their sustained popularity.

The jury is still out on the mass-production of propane and natural gas powered vehicles. While they burn fuel very cleanly and efficiently, they still do not have the performance of the gasoline or even diesel powered vehicles.

Our Environment

Now, about the environment. Did you know our air is as clean as it was in 1970? That's great, right? Maybe. For those who remember the 1970's, that era saw the last of the car manufacturer horsepower wars. Loose emissions regulations for light vehicles allowed much higher levels of pollutants from vehicle exhausts. So, why isn't our air more clean now than it was then? It would appear we have not cleaned up our act very much.

Actually, our vehicles are much cleaner than they were back in the 1970's, it's just that we have 150% more vehicles on the road today than we had back then! If you choose to look at it from that perspective, our per vehicle pollution level has decreased significantly.

Alternative fuels such as E85 and bio-diesel are significantly less harmful to the environment, emitting fewer greenhouse gases and other pollutants. In the future, fuel cell technology promises to emit only water vapor!

If interested, you can get a two-day clean air forecast from the EPA at: www.airnow.gov

Make sure to read the section called "Pollution and the Role of the Catalytic Converter" in Chapter 4. It goes into much more detail about the specific pollutants and greenhouse gases being emitted from our automobiles.

Wouldn't it be great one day, when you know the millions of automobiles carrying people to their destinations, and the hundreds of thousands of semi-tractor trailers delivering goods for this country are no longer putting harmful substances into the air? We think so.

Chapter 6: Driving Technique and Energy Management

[D]

Driving Technique Can Improve Your Gains or Take Them Away!

Let's review the three factors affecting your vehicle's fuel economy and performance. Remember LED?

We have looked at the sources of energy **loss** in vehicles and maybe you have already followed up on some of the solutions we've presented. Losses are the forces that take away from the total amount of energy available to your vehicle.

We've also looked at engine **efficiency**. We learned we could squeeze an optimum amount of efficiency (and energy) from the internal combustion engine. The limits to this efficiency are based on the physical capabilities of the engine itself, the fuel used, and the regulatory environment to which the automakers and we must adhere.

Beyond the optimization of losses and efficiency, your **driving** habits and environment are the only remaining factor affecting your fuel economy. We believe our driving habits account for a substantial of the overall fuel economy we can hope to achieve with our Gas Guzzlers!

Another point we'd like to make right here is most of the time, when drivers see an improvement in fuel economy and performance, they know in the back of their minds their vehicle is operating much more efficiently.

They think (or, more appropriately, they DON'T think) it won't hurt to jam on the accelerator a bit. I mean, it is fun to feel that added performance, and after all, they do have a more fuel efficient vehicle, right? What's the harm in getting on it a little bit?

Well, the harm is this: getting on it will ruin all the gains you've painstakingly made by eliminating losses and making your engine more efficient. Your driving habits can improve on the gains you've made, or they can take them away!

We're pretty sure that when we see negative press regarding a device or technique we know from experience works, it wasn't the device or technique at fault. We're pretty sure it was the driver at fault.

Consciously or not, the driver probably enjoyed the new-found performance associated with fuel mileage optimization a little too much. Let's face it, its fun to jam on the accelerator and feel that big ole' beast MOVE!

Much of the time we are not even conscious of the fact we are being aggressive on the gas, especially after the vehicle has undergone a transformation and it is suddenly much more effortless to move from point A to point B.

We are asking you to be aware of this phenomenon we call *subconscious passive aggressive fuel economy psychosis* (our contribution to the field of psychology!) and use the following driving techniques to further extend the gains you've already made.

Energy Management

Energy management is the single most important concept we will introduce to you. Actually, you've been introduced to it before, but probably haven't considered it in these terms. So, what do we mean by managing energy?

In high school physics, you may remember that energy can not be destroyed; it can only be converted to different forms. And remember Newton? The guy with the apple falling on his head? His laws explain how much of the physical world operates. He had some funny rules regarding objects in motion, and equal and opposite reactions. You remember the guy!

We don't need to go into all the techno-babble behind this, but just remember that what we are going to discuss relative to driving technique and energy management is grounded in many of these principles.

Conservation of energy is the key to good driving habits resulting in good fuel economy. It takes a certain amount of energy to get your car moving from a dead stop to cruising speed (called acceleration). It takes much less energy to keep the car moving at a steady speed. Every time we must slow down or stop, we need to accelerate again to get the vehicle up to cruising speed.

Let's discuss the forms energy takes during this scenario.

To accelerate from a dead stop, we press on the gas. This initiates a series of actions resulting in fuel being burned in our engine and a chemical reaction that converts the energy in gasoline (ok, only 20% of it) into mechanical energy which moves the car forward.

We know from previous chapters that we lose quite a bit of energy along the way to friction, mechanical load, heat and vibrational noise. Remember, it is not actually lost, but converted to heat and other forms of energy.

But, the engine generates enough energy so that it can afford to lose some and still get the car moving. Once the vehicle is up to speed, we can let off on the gas a bit and let it cruise. The energy from the gasoline is converted to mechanical energy, which is converted to kinetic energy (remember that from high school?).

Kinetic energy is the energy contained in an object with momentum. You can let off the gas and the vehicle continues to roll for a while, slowly converting its kinetic energy to other forms of energy due primarily to rolling resistance and other forms of friction. When you let off the gas and step on the brake pedal, energy is converted very quickly to heat via friction of the brake pads against the rotors or drums of your vehicle.

We will say, then, that when we step on the brakes, we very quickly bleed off or lose our kinetic energy (yes, we know it is not destroyed, just converted to heat, which is useless to us) and to get back up to cruising speed we must start all over again in building up momentum.

Unfortunately, that high energy state we were in, achieved by using up fuel, was thrown out the window by the simple act of hitting the brakes. We know that using the brakes is important in the day to day business of driving our vehicle, and we can't do without them.

The Gas Mileage Bible

Trading Altitude for Airspeed...

Fighter jet pilots learn all about kinetic and potential energy as a means of mastering their dog-fighting skills. He (or she!) who has the most energy at the end of a dog-fight is the one to "pull to guns" on that enemy fighter – the winner.

Even the most sophisticated Navy jet with massive afterburners has a fixed amount of energy. Generally speaking, that energy is the sum of the energy contained in its fuel, the ability of the engines to convert the fuel to forward thrust, the kinetic energy of the jet (airspeed), and the potential energy of the jet (altitude).

Every maneuver, every pull on the stick uses up some of that energy. Getting on the 'six of that enemy fighter requires a balance of all the forms of energy available to the pilot. Add some thrust here; turn harder there.

Frequently, pilots need to gain airspeed very quickly – faster than their engines will accelerate the jet. So, they trade altitude for airspeed. They go into a zero G force dive that lets Mother Nature (gravity) help the jet accelerate. They trade potential energy for kinetic energy. Get it?

Similarly, when jets are flying a mission very close to the ground at high speeds, they use the reverse situation if they get into trouble. You don't want to troubleshoot a "Check Engine" light flying so fast and so low, so you pull the nose up and trade airspeed for altitude – you want a little breathing room, see? You are trading kinetic energy for potential energy.

continued on next page...

> **Trading Altitude for Airspeed... Continued**
>
> We are going to learn some of the same techniques for harnessing kinetic and potential energy while driving our gas guzzler. We'll learn to trade one type of energy for another when feasible, and learn to maximize our energy state at all times.
>
> And, NO, we are not going drive real fast, fly upside down, and shout "I have the need, the need for speed!" That's in someone else's book (or was it a movie?).

But, since we want to conserve our energy, and braking converts our useful energy to useless energy, *could we manage our driving by braking less? Or at least braking less aggressively?*

Same thing with acceleration. Jamming on the gas to speed up quickly is a notorious energy waster. The engine consumes vast amounts of air and fuel during a hard acceleration, but it is not operating as efficiently during that time. The fuel does not burn as completely in large doses as it does in smaller doses. The engine must also overcome considerable mechanical load and friction very quickly.

All this adds up to a very expensive five or ten seconds worth of holding your gas pedal down while you get up to speed.

Let's summarize what we know so far about conserving energy:

- Having kinetic energy (momentum) is a prized energy state

- It takes more energy to accelerate than it does to maintain a constant speed
- It takes far more energy to accelerate quickly than it does to accelerate at a moderate rate.
- Braking throws all your hard earned momentum out the window, and we must start over again, trading fuel energy for kinetic energy

Now let's look at some driving scenarios and see how all this applies to saving gas.

Bad Driving Behavior

We have all been guilty of the following driving scenarios:

Scenario #1   Screech and Roar

Driving as fast as you can, maintaining speed until the last possible moment before a stop sign, at which time you hit the brakes hard, decelerating rapidly and coming to a stop with your tires barely holding the road. Then punch the gas to accelerate to cruising speed again. Repeat again and again until you are on time for work.

Scenario #2   Pushing the Envelope

Driving right on someone's tail, trying to push him or her to go faster or position yourself for the big passing move. You find yourself accelerating to a position just a few feet behind the other vehicle, and then realizing how close you are, let off the gas and drift back to a safer following distance.

Before you know it, your subconscious foot has accelerated you to within a few feet of the car ahead of you again.

Repeat until they go faster or until you've arrived at your destination.

Scenario #3   No Man Shall Pass

Driving in bumper to bumper highway traffic, you are zealously guarding your position in line. You couldn't in good conscience possibly allow someone to bully his or her way in front of you.

So, you're working the gas and the brake, keeping right on the tail of the car in front of you, not leaving room for anyone else to sneak in. You see someone making a move on the inside lane, so you jam on the gas and move it in a little closer, just to be sure.

Scenario #4   Kick Back and Cruise

You have a long drive ahead of you. It happens to be over some rolling hills. You had a rough night and need a little rest, so you hit the cruise control and sit back to enjoy the ride.

You let the cruise control manage your car's energy up and down those hills, all the way to your destination.

Scenario #5   Techno-Rabbit

Electronic warning systems flashing and beeping, both hands gripping the wheel, head pushed forward with eyes darting back and forth, looking for the enemy state trooper you know is out there. Flying past traffic, ducking left and ducking right, you're only doing 95 right now, but wait till the road ahead clears up again! If all goes well you should get there before you left.

Ok, so maybe a little exaggeration with these scenarios. But maybe not? If we haven't lived these scenarios, we certainly have seen others act this way.

This book is not going to single handedly change the way the human race drives. That's because human driving behavior is modeled after human behavior. Period. If anything, we are more aggressive in our vehicles because we have a sense of protection and isolation. We don't see the effect of our driving behavior on others – at least not all the time. It is not up close and personal when we are behind the metal and glass of our vehicles.

This book won't automatically change your driving behavior. It won't make you leave sooner so you don't need to drive as fast. It won't change how you feel about looking cool or letting people cut in front of you. These are behavioral changes you should work on.

But, maybe, this book can help open your eyes so you can be aware of how your personal motivations, habits and hang-ups affect your driving technique. And how your driving technique affects your gas mileage. And how your gas mileage affects your pocket book. And how, if gas prices keep going up, maybe your pocketbook will affect your behavior…

Good Driving Habits

So let's re-examine the driving scenarios listed above using good energy conserving driving habits:

Scenario #1   Screech and Roar *becomes* Steady and Coast

Accelerate slowly and steadily until you reach your cruising speed. Anticipate having to slow down or stop and let off the gas well ahead of time. Let your momentum carry you forward, trading momentum for distance.

Don't hesitate to use the brakes any time you need them, but try to use them minimally.

Let natural rolling and wind resistance slow you down, letting your momentum carry you as far as possible, and use your brakes lightly to come to a final stop.

Scenario #2   Pushing the Envelope *becomes* Pace Setter

Leave early enough to drive at a leisurely pace. On flat terrain, set your cruise control to drive 55 or 60 miles an hour where you will get dramatically improved gas mileage. Stay in the right hand lane unless passing someone. Let the rat race and associated driving tension pass you by.

You'd be amazed at the difference in your stress level when you're not playing Mario Andretti on the drive to and from work.

Surprisingly, it won't take you much longer, and you'll have time to think, listen to good music or listen to an audio book on the way. And when you get home you may not even feel like kicking the dog!

Scenario #3 No Man Shall Pass *becomes* Make a Hole

Rather than fighting your way to the front of the pack, lay back and stay in the right hand lane. Create a nice wide gap between you and the vehicle ahead of you. Maintain steady pressure on the gas pedal.

If the cars ahead of you slow down, use the gap ahead of you to slow down gently by coasting, only using the brakes if necessary. Once the speed of the cars ahead of you picks up again, accelerate gently, reestablishing the gap in front of you.

You will find keeping your gap and not driving five feet from the next person's bumper is infinitely less stressful. Again, amazingly, you will still arrive at your destination, stress free, not long after the speed demons who fought and scratched for position all the way there.

Scenario #4 Kick Back and Cruise *becomes* Speed Down and Coast Up

Cruise control is not a friend of fuel economy on hills. It doesn't know when the next hill is coming and doesn't know some of our secrets for taking advantage of gravity. It doesn't understand that accelerating up a steep hill makes for horrible gas mileage.

Here's a better way. Keep the cruise control turned off when driving on rolling hills. As you are headed down a hill and ready to start up the next one, accelerate gently. Use the force of gravity to assist your engine in the acceleration process. Accelerate up to a speed you are comfortable with, but not too fast. If you constantly accelerate to 70 or 80 miles per hour down the hill, you are probably not gaining much since your fuel burn rate is so much higher at higher speeds. And you are probably in line for a traffic ticket from your friendly state trooper.

As you start climbing the next hill, watch your speedometer and ever so slowly let the gas off. Let the momentum of the car carry it up the hill with a minimum of pressure on the gas pedal, letting the pressure off slowly. Let the car slow as it nears the top of the hill, but no slower than a few miles per hour slower than the speed limit.

Once you reach the crest of the hill and begin down hill, start the process over again. Do not use this technique when there is traffic right in front of or behind you, and stay within a reasonable range of the current speed limit – not too fast and not too slow.

The idea is to take advantage of gravity to accelerate down the hill, and then convert your kinetic energy to distance traveled (trade speed for distance) on the way up the next hill.

Scenario #5   Techno-Rabbit *becomes* Steady Traveler

While driving at 50 or 55 miles per hour may be boring to some, driving at that speed can save as much as 23% of fuel cost according to the Department of Energy and the Environmental Protection Agency (EPA).

The EPA claims that for each 5 mph over 60 mph, it is like paying an additional $0.21 per gallon for gas. Or, another way to look at it, for every 10% faster you go over 50 mph requires an additional 10 to 20% of fuel.

For example, if you are getting 15 mpg at 50 mph, you could expect to get from 12.5 to 13.6 mpg at 55 mph. This is a rule of thumb, because of the variations in vehicles, drivers and environments.

Here is a graph from the EPA showing average fuel economy at varying speeds. See how quickly the mileage falls off after 50 or 55?

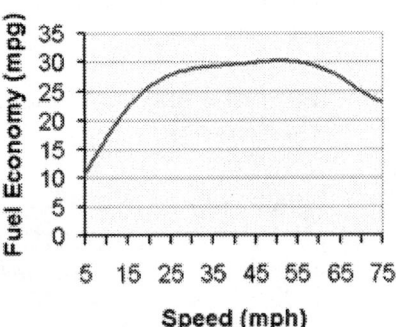

If you get in the habit of leaving in plenty of time, you won't need to rush and you can take advantage of the slower speed to save you big money.

Again, this may be a complete lifestyle change for some of you. Many people live life in the fast lane. You drive fast, you eat fast, you leave at the last minute, everywhere you go and everything you do. That's ok. That's your choice.

We, the authors, have chosen to slow down, create more balance in our lives and not live life in the fast lane. We've compared notes and find that driving in a more leisurely manner has not taken any time out of our lives. On the other hand, we somehow feel we've gained more time. We both feel like we have more time in the vehicle to listen to good music or an audio program. And, maybe, since we must plan a little better to leave on time, it seems other areas in our lives are working better as well.

We arrive stress free and without white knuckles. It doesn't feel like we leave any sooner than we did before when we used to race everywhere we went. It's as if we were given time back in our lives. All that from slowing down!

Bottom line – you can choose to say driving slower is boring, or you can choose to say it gives you more time for yourself. Either way you'd be right. But only one way you'll be saving fuel and money!

Let's now summarize the driving tips and techniques we've covered so far:

Solution #33   Accelerate steadily, but not aggressively to get to cruising speed

Aggressive acceleration is an energy consuming activity. Unfortunately, most people do it out of habit. We must accelerate, or we'll never start moving. A slower and steady acceleration is the least energy consuming. Try it once. Accelerate steadily out of a red light – not slow but not fast. You'll see people on both sides, and ahead of you pulling away. The people behind you may want to quickly go around you. But, if you can establish this habit, it will pay great dividends in terms of saving money at the pump.

One way to think about it is to pretend there is an egg under your gas pedal. If you press too hard and too fast, you break it.

The Gas Mileage Bible

Manage your rev's...

Remember, the more RPM's your engine generates the more fuel you burn. Try this. Look to see what RPM reading you have at 60 miles per hour. Use that number as a guide to manage your accelerations while driving in town.

For example, if your vehicle runs at 2,000 RPM when you are driving at 60 MPH, then use 2,000 as your guide number. Every time you accelerate, watch the RPM out of the corner of your eye, and see if you can get up to speed without exceeding your guide number (2,000 in this example).

You'll be surprised at how gently you need to press on the accelerator to stay below your guide number. You'll also be amazed at all the people flying by you. If you look closely, you should be able to see little dollar bills flying out of their tail pipes. ☺

Just accelerate smoothly (another good rule of thumb is to accelerate no more than 3 mph per second) from each stop in the city while never exceeding your MAX RPM at, say, 60 MPH. Kenny has been able to get over 17 MPG in city driving in the rolling hills of Austin, Texas with his Suburban following this technique.

Solution #34 Anticipate stops or slow-downs and minimize use of your brakes

Pay attention to your driving situation. Leave plenty of room in front of you, and place your vehicle where you can see what's happening ahead. Watch for brake lights coming on, traffic lights preparing to change, cars pulling out into traffic – anything at all that may cause a slow down. Scan the area just in front of your vehicle as well as far ahead so assess the situation.

When you see a stopping or slow-down situation developing, let off the gas and let the natural air resistance and rolling resistance slow the car. You are trading speed for distance here and not throwing hard-earned energy out the window.

As we mentioned before, using your brakes instantly converts your precious kinetic energy, or forward motion, into heat, something wasted in most cars. Only some newer hybrid cars have regenerative braking systems to recapture some of that energy.

Never hesitate to use your brakes when necessary, but if you learn to anticipate stops and slow-downs, you can use them less. Again, this technique requires *awareness*. If you are one who does not pay attention while driving, this technique is not for you.

Solution #35 **Leave early enough that you don't feel like you need to rush to get there in time**

This is more a choice you make more as a lifestyle habit than anything else. You may not be able to resist the urge to leave at the last minute and then race to work or school. If you can (and we suggest you do), leave early, drive slower and arrive fresh and stress free. Try it, you may like it.

Solution #36 **Drive slower - no faster than 55 or 60 mph if feasible**

There are plenty of tests out there to show that driving slower saves gas. Remember the aerodynamics lesson? It takes *8 times the power* to overcome *4 times the drag* when you *double your speed*. Keep the speed, the drag and the power requirements lower!

Of course, if you are in a situation that can become dangerous by driving slower than the rest of traffic, use your judgment and be safe.

Refer back to pages 87-88 for more on how much driving faster really costs you.

Solution #37 **Don't tailgate – create a safety "bubble"**

Keep a gap between you and the car in front, just like the Driver's Ed manual says. If you drive too close, you are constantly adjusting your speed with the gas pedal, wasting a bunch of gas.

The bubble of space you create protects you in case traffic comes to a sudden halt ahead. It lets you keep a steady pressure on the gas pedal even while the crazies in front of you are speeding up and slowing down like a yo-yo. Remember the three second rule? It should take you three seconds to get to a landmark being passed by the car in front of you.

Here's how it works if you've not heard of this handy rule of thumb. Find a landmark like a mile marker or street sign and count how many seconds it takes for the landmark to pass from the back bumper of the car in front of you to the front bumper of your vehicle. You will be surprised at how far you must back-off the gas pedal to stay 3 seconds behind the person in front. It gives you substantially more time to react and slow down.

In the absence of an emergency stop ahead, you can use the gap to decelerate slowly when traffic ahead slows down, trading speed for distance as you close the gap slowly. When the pace picks up again, rebuild your gap by accelerating slowly and steadily.

Solution #38   Drive in the right lane except when passing

Don't try to protect your position in line. Let the rabbits get past you. Keep your gap (the empty space ahead of you) and use it to maintain momentum, to buffer changes in speed of the traffic around you, and to minimize accelerations and braking.

Solution #39   Don't let your vehicle idle for long periods of time

As a rule, if you think you will be stopped and idling for more than a minute, put the vehicle in Park, and stop the engine. Use your common sense apply this recommendation appropriately. Don't stop your engine when stuck in stop and go traffic, for example.

When you stop at the mini-mart to run in for a gallon of milk or a six-pack, shut your car or truck off. Those five minutes of idling will kill your mileage.

On those cold mornings, try to limit the amount of time the car idles while warming up. 15-45 seconds is PLENTY of time for the engine to get lubricated. This is the absolute worst time for wasted fuel energy and for excessive emissions...

Remember, your catalytic converter doesn't start cleaning the exhaust air until it is warmed up. The engine and the catalytic converter warm up faster when you are driving, and take longer to warm when the vehicle is just idling.

Solution #40   Drive with only one foot

Have you ever driven behind someone with his brake lights on all the time? Even when driving down the highway? You know what causes that don't you? Driving with two feet

– one on the gas and one on the brake. When you drive with two feet, it's easy to get in the habit of resting your left foot on the brake pedal.

The problem with driving with two feet, other than confusing everyone driving behind you, is that your brakes are probable being applied ever so slightly, causing lots of heat, friction and extra gas to be burned to keep your car moving.

Most of us learned to drive with only one foot, but a few people still insist on driving with two. If you are one of those people, please be aware of what your left foot is doing at all times!

Solution #41   Use your cruise control on the flats

Using your cruise control on long, flat stretches of road can help you save gas by maintaining a consistent speed. Lots of us unknowingly speed up and slow down slightly when we are trying to maintain a steady speed.

It is human nature for our attention to wander, and when that happens, we may end up letting off the pressure or increasing the pressure on the gas pedal unknowingly. Then our attention comes back into focus and we correct our speed, sometimes aggressively, which unnecessarily used more fuel.

We do NOT recommend you use cruise control when the road you are traveling has hills or steep grades. The cruise control computer has no way of anticipating the next big hill or know when you are going *down hill*. It does not take advantage of downhill momentum being generated. On a steep uphill stretch, you will hear and feel the engine racing and maybe even shifting into a lower gear to get to the top. And then, as soon as you crest the hill, the cruise control will let off on the accelerator and start coasting down the hill.

This where we recommend, to the extent it is safe and within traffic laws, using the next technique.

Solution #42   Let gravity help you on hills

Refer back to Scenario #4 on page 87. This technique involves a little practice and focus but it can make a huge difference in your gas mileage if you drive on rolling hills.

The technique can be summarized as follows, but refer back to the chapter to learn the fine points:

- Accelerate or coast gently down the first hill, letting gravity take some of the workload off your engine, and building up a little excess speed

- Continue to accelerate gently as you get to the bottom of the hill and start up the next one

- Part way up the hill, start letting off the gas pedal, ever so slightly, allowing the car begin losing a little speed

- As you near the top of the hill, let more pressure off the gas pedal, allowing your car to continue slowing gently as you climb

- By the time you have reached the top of the hill again, you are back to the speed you started at, and your engine has worked a whole lot less in getting you there

The idea is to let gravity help accelerate you down the hill, building a little excess speed, which you then bleed off as you start up the hill.

Your engine has to work much less going up the hill, because you have gotten a "running start" so to speak, and you are letting the car "coast" up the hill with only a little assistance from the engine.

Again, please use common sense and obey all traffic laws. Don't try this technique on a busy road with lots of traffic ahead of and behind you.

Solution #43 **Watch your fuel economy monitor**

Many newer cars have the ability to monitor fuel economy while driving. This often includes the ability to watch *average* fuel economy as well as *instantaneous* fuel economy. Average fuel economy is just that, the average of miles per gallon over a period of time, including all the driving you've done from the time you reset the monitor.

Instantaneous fuel economy shows how much fuel you are using at any given point in time. For instance, it might show you are getting eight miles per gallon while accelerating, 15 miles per gallon while cruising along at 60 miles per hour, and 25 miles per gallon while you are slowing down. Over time, all these will average out to what your average fuel economy reading says.

The great thing about a fuel economy monitor is you can see the effect of your driving habits instantly and use the feedback to change how you drive to get the best possible mileage. You will be amazed to see how awful your instantaneous gas mileage is at certain times, like when accelerating or when climbing a hill. You'll be equally amazed at how easy it is to make subtle, minor adjustments to your driving style leading to big gas savings.

Solution #44   Install a "poor man's" fuel economy monitor

Installing a vacuum gauge on or under your dashboard will help you monitor your fuel use. The reduced pressure in the intake manifold caused by the pistons trying to draw air is called vacuum. The vacuum pressure is high at idle or low load conditions and high at full throttle or heavy load conditions.

While the vacuum gauge doesn't tell you what your fuel economy is, it helps you monitor how much fuel you are using by how much pressure there is in the intake manifold. If you see the gauge constantly at a "low" reading, you may be standing on the accelerator a bit too much. You can reduce your aggressive driving habits by watching the relative position of the vacuum gauge and trying to maintain the highest reading you can under various driving conditions.

Solution #45   Get in a conservation state of mind

Your daily driving habits are just that – habits. New habits can be formed. Bad habits can be changed. To really get the benefit of optimum gas mileage and money savings you will want to get into a conservation state of mind.

This state of mind helps you set good driving habits, and helps you with the discipline of keeping a mileage and maintenance log. It helps you decide if you really need to make your midnight run to the store for ice cream. And, it will help you decide to walk or ride the bike on those nice days.

The attainment of a conservation state of mind is a direct extension of how you manage the rest of your life. This makes sense for many of us who are seeking more balance in our lives and are perhaps looking for ways to slow down the pace a bit.

Taking the perspective that your daily commute can be a peaceful and stress-free event where you get some "me time" is a choice you make.

Look at it this way. If you truly believe your commute will be a daily journey through hell with no chance of ever being peaceful – then it will be. Hell that is.

If you truly believe, even in the worst of traffic nightmares, you can find a time of quiet and solitude for a while, to enjoy a good audio book, or to listen to some great music, then you will find what you are looking for.

Use the opportunity to drive a little slower, take on some new and improved driving habits, save some money and enjoy yourself more. It's only a choice away.

Chapter 7: Putting It All Together

Recap - LED Method for Getting Better Gas

We all love our gas guzzlers. Or, we love to hate them, but many of us don't have a choice about owning and driving one. Either way, we want to find a way to pay less for the privilege of driving them. We (the authors) hope you found the information in this book helpful. We're confident that, based on many years of our own experience, you can very easily get a 10%, 20%, 30% or more improvement in gas mileage using only a few of the tips we've presented here, and no gimmicks. No chemicals, gas pills or gadgets.

We can't tell you exactly how much or how little each of these tips and techniques will help your vehicle in your situation. We know from experience the cumulative effect of just a few of these techniques will get you to the level of improvement stated above. It will require your new found awareness, diligent focus and your intention to be successful. It will require shifting the way you think about some things. And, it will require you to be disciplined. Before you know it, new habits will be formed and you will hardly even notice you are driving more carefully and doing all the right things to really put some gas money back in your pocket.

Remember, LED! Cut the *Losses*, improve the *Efficiencies* and be aware of your *Driving* habits. Following these three simple guidelines could save you $1,000 in gas in the coming year, especially with the skyrocketing price of fuel.

Losses include all the forces taking energy away from us. Types of losses include friction, mechanical load and heat/loss. Other physical forces can also be sources of loss to our vehicle. For example, gravity can be an energy draining force if we are going uphill (or a helping force if we are going down). There are many ways for losses to creep in and reduce our fuel economy and performance. Old oil, under-inflated tires, front end mis-alignment, engine out of tune, improperly working sensors. These and many other maintenance problems on our vehicle can cause significant energy loss directly proportional to our gas mileage and performance.

Efficiency lets us convert the maximum amount of energy from our fuel and convert it to mechanical energy, or forward momentum of our gas guzzler. There are many nuances to the fine tuning of our engine. Getting all the engine subsystems to work together to perform an exquisite symphony under the hood of our car takes effort and discipline on our part. Fuel quality and type plays an important part in the efficiency formula since it is the source of the energy propellng our vehicles forward.

Driving technique is perhaps the most important of the three aspects of saving gas. It is the most important because your driving technique can quickly *destroy any gains* you have made by eliminating losses and improving efficiency. Conversely, your driving technique alone, or with improvements in losses and efficiency, can significantly improve your gas mileage.

Getting in a conservation state of mind is important to adopting good driving habits. Being a successful conservative driver may require other lifestyle changes as well. Learning to leave earlier for work or appointments, not being in such a rush, slowing things down a bit. These are all traits we think are important to implementing good driving habits. Otherwise, you will still be driving fast, tailgating and pretending the road to work is the Indy 500 track.

The Psychology of Energy Management

Energy Management is a state of mind. It means getting in a conservation state of mind. Once you are thinking that way you can begin to form habits. Good habits. Habits that will help you drive more efficiently and safely without you having to think about it all the time.

It also means being aware of how your vehicle converts energy from fuel to motion, and what you can do to help the process along. The psychology of energy management means that you are aware of your driving conditions and environment. If you are driving on flat, open highways, you know to use your cruise control. If you are driving on rolling hills, you know how to use gravity to your advantage and improve your mileage. If you are in slow, bumper-to-bumper traffic, you know to create a bubble in front of you. You know how to use your space to create a safety barrier, and to minimize acceleration and braking.

Saving gas is a state of mind. Yes, it also requires maintenance and some elbow grease. But it is mostly a state of mind. A state of mind that is with you every time you get in the car and buckle up. It is your state of mind when you accelerate from a stop light, and it's your state of mind when you're getting ready for work in the morning – preparing to leave early so you need not rush. It is your state of mind when you log your mileage every time you fill up; and when you inspect your car for ways to eliminate losses and improve efficiency.

The psychology of energy management can even be extended to the way you manage yourself. Live your life in balance. Reduce stress. Stop rushing through life. Stop and smell the roses. Save your energy for what's important, not for yelling and gesturing at fellow drivers on the way to work.

> **Mental Fatigue for the Jet Jock...**
>
> Flying a combat mission in a modern jet aircraft is possibly one of the most tiring activities a human can endure. Attack aircraft generally fly very low to the ground at high speeds on the way to their targets to avoid radar and being spotted. It takes less than a half second of lost concentration during a low altitude, high speed maneuver to turn that expensive jet into a smoking hole in the ground. And no, you won't have had time to eject.
>
> Between holding altitude, watching for obstacles, and successfully maneuvering the aircraft, the aircrew must listen to and talk on several different radio frequencies at once, navigate to the target, prepare the weapons for launch, monitor all the aircraft systems, and oh yeah – they must watch out for enemy fighters, missile sites, and unforgiving radio towers or mountain peaks.
>
> A three or four hour mission can wipe you out for the rest of the day. Extremely exhausting!
>
> So, why would you as a driver put yourself through the same thing on the way to work?
>
> continued next page...

The Gas Mileage Bible

Mental Fatigue for the Jet Jock... Continued

Consider this - you are in heavy highway traffic. You tailgate the person in front of you with only a few feet between your two vehicles. You rapidly and alternately mash on the gas and the brakes to maintain your distance. You don't want to hit the guy, but you don't want to let that bogey slide in from the other lane either. Your concentration is intense. You have sweat on your brow and down the center of your back. Whoa! Sudden stop ahead. You jam on the brakes and consider, just for a moment, taking the grassy median to prevent the collision. No – you've got it. Inches to spare. OK, everyone is accelerating again, but you're tight on the guy ahead. No way are you letting anyone cut in.

How much personal energy does that take? We know from having been there, that it takes a lot of energy. Why show up for work at the beginning of the day already mentally worn out and dripping with sweat?

Don't be a jet jock. Get out of the fast lane. Slow down and relax. You may just like it! And, you may just avoid that smoking hole.

Kenny Joines & Ron Hollenbeck

Part Three Additional Information and Resources

Factors Affecting Gas Mileage (Other than the Energy Content of Fuel)

Following is a table pulled directly from the EPA website (www.epa.gov) showing the potential impact of various factors (environmental, vehicle condition, and driver technique) on fuel performance.

Effect	Conditions	Average Fuel Economy Reduction	Maximum Fuel Economy Reduction
Temperature*	20F vs 77F	5.3%	13%
Head Wind	20 mph	2.3%	6%
Hills/Mountains	7% road grade	1.9%	25%
Poor road conditions*	Gravel, curves, slush, snow, etc.	4.3%	50%
Traffic Congestion	20 vs 27 mph average speed	10.6%	15%
Highway speed	70 vs 55 mph	N/A	25%
Acceleration Rate	"Hard" vs "Easy"	11.8%	20%
Wheel Alignment	1/2 inch	<1%	10%

Tire Type	non-radial vs radial	<1%	4%
Tire Pressure*	15 psi vs 26 psi	3.3%	6%
Air Conditioning	Extreme Heat	21%	N/A
Defroster*	Extreme Use	Analogous to A/C on some vehicles	
Idling/Warmup*	Winter vs Summer	Variable with Driver	20%
Windows	Open vs Closed	Unknown but likely small	

Learn more about saving money and getting better gas mileage

If you would like to continue learning more about saving gas and improving the performance of your vehicle, you can sign up for our free newsletter and a free email course where we feature additional tips and techniques for saving gas:

www.gasmileagebible.com

Each newsletter includes one or more of the following features which may be of interest to you:

- additional tips and techniques for saving gas
- recommended products for improving fuel economy and/or performance
- updates on new technologies
- newsworthy events related to improving fuel economy or performance
- cool new products for your gas guzzler
- special offers for new or related products or services

Other books and products available

Watch for more valuable books and products at the following website:

www.vitalshift.com

Part Four   Index of Tips and Techniques

Tips for Reducing Losses

Solution #1	Keep your tires inflated properly..	45
Solution #2	Don't drive with your windows down	46
Solution #3	Apply synthetic bearing grease to your wheel bearings	46
Solution #4	Use synthetic oil in your transmission, differential and transfer case	47
Solution #5	Make sure your wheels are aligned to reduce rolling resistance..	49
Solution #6	Maintain your brakes and make sure your emergency brake is disengaged	50
Solution #7	Make your vehicle more aerodynamic................................	51
Solution #8	Remove all unnecessary baggage and payload from your vehicle	64
Solution #9	Keep your air conditioning turned off when it is not hot........	65
Solution #10	Keep unnecessary electrical accessories turned off	66
Solution #11	Use high quality synthetic oil in your engine	66
Solution #12	Consider installing an electric radiator cooling fan	69
Solution #13	Keep your fuel filter clean..	69
Solution #14	Ensure your power steering fluid and belts are properly maintained	70
Solution #15	Make sure you are using the proper coolant in your vehicle...	70
Solution #16	Clean your engine at least once a year................................	73

Tips for Increasing Efficiency

Solution #17	Replace your factory paper air filter with a High Flow Replacement Air Filter	81
Solution #18	Install an Air Vortex Device	82
Solution #19	Install a Throttle Body or Manifold Spacer	84
Solution #20	Clean and Polish Your Mass Airflow Sensor (MAS)	86
Solution #21	Install an Air Intake System	86
Solution #22	Clean or replace your fuel filter regularly	91
Solution #23	Buy the lowest octane grade of fuel recommended for your vehicle	92
Solution #24	Check your gas cap	93
Solution #25	Don't let fuel get stale	94
Solution #26	Clean and service fuel injectors as recommended by your service manual	96
Solution #27	Use high quality spark plugs and ignition cables	106
Solution #28	Keep your engine tuned and well maintained	108
Solution #29	Tune your car up and get an exhaust gas analysis	118
Solution #30	Make sure your Oxygen Sensor is Working Properly	119
Solution #31	Install Lower Resistance Exhaust Components	120
Solution #32	Follow a regular maintenance schedule	126

Tips for Driving Technique

Solution #33	Accelerate steadily, but not aggressively to get to cruising speed	148
Solution #34	Anticipate stops or slow-downs and minimize use of your brakes	149
Solution #35	Leave early enough that you don't feel like you need to rush to get there in time	150
Solution #36	Drive slower - no faster than 55 or 60 mph if feasible	151
Solution #37	Don't tailgate – create a safety "bubble"	151
Solution #38	Drive in the right lane except when passing	152
Solution #39	Don't let your vehicle idle for long periods of time	153
Solution #40	Drive with only one foot	153
Solution #41	Use your cruise control on the flats	155
Solution #42	Let gravity help you on hills	156
Solution #43	Watch your fuel economy monitor	157
Solution #44	Install a "poor man's" fuel economy monitor	158
Solution #45	Get in a conservation state of mind	159